Coronaviruses Research in BRICS Countries

Coronaviruses Research in BRICS Countries

Editors

Burtram C. Fielding
Georgia Schäfer

MDPI • Basel • Beijing • Wuhan • Barcelona • Belgrade • Manchester • Tokyo • Cluj • Tianjin

Editors
Burtram C. Fielding
Medical Biosciences
University of the Western Cape
Bellville
South Africa

Georgia Schäfer
Virology – Emerging Viruses
ICGEB Cape Town
Cape Town
South Africa

Editorial Office
MDPI
St. Alban-Anlage 66
4052 Basel, Switzerland

This is a reprint of articles from the Special Issue published online in the open access journal *Viruses* (ISSN 1999-4915) (available at: www.mdpi.com/journal/viruses/special_issues/SARS_BRICS).

For citation purposes, cite each article independently as indicated on the article page online and as indicated below:

LastName, A.A.; LastName, B.B.; LastName, C.C. Article Title. *Journal Name* **Year**, *Volume Number*, Page Range.

ISBN 978-3-0365-4028-3 (Hbk)
ISBN 978-3-0365-4027-6 (PDF)

Contents

About the Editors

Burtram C. Fielding

Professor Burtram C Fielding is currently the Dean of the Faculty of Natural Sciences at the University of the Western Cape. As a researcher, Professor Fielding is the Principal Investigator for the Molecular Biology and Virology Research Group in the Department of Medical Biosciences at UWC. A UWC alumnus—having completed all his undergraduate and postgraduate studies, including his doctoral work, at the mentioned university—Professor Fielding's research focus has been on applied molecular biology in human health. Specifically, his main area of research is the molecular virology of human coronaviruses, although his lab is also involved in studying the molecular biology of medicinal plants as potential treatments for human diseases. Burtram Fielding has been studying coronaviruses for some 20 years, starting in Singapore—where he was a Research Fellow in the Collaborative Anti-Viral Group at the acclaimed Institute of Molecular and Cell Biology (IMCB)—with SARS coronavirus in 2003. A fervent believer that scholars have a role to play in society as public intellectuals, Professor Fielding has a passion for sharing COVID- and SARS-CoV-2-related research with the public. To this end, he has published several opinion pieces in the popular press; has been interviewed and quoted in countless international magazines; and has been interviewed on a host of continental and international television and radio channels. He conducted hundreds of radio, television and print interviews during the COVID-19 pandemic, appearing in prominent local and international media, in the Middle East, China, and other Asian countries, often as the sole voice from Africa and the Global South.

Georgia Schäfer

Dr Georgia Schäfer is the Group leader of the Virology-Emerging Viruses Group at the International Center for Genetic Engineering and Biotechnology (ICGEB) Cape Town and a Senior Fellow at the European and Developing Countries Clinical Trials Partnership (EDCTP). She is also a Full Member of the Institute of Infectious Disease and Molecular Medicine (IDM) in the Faculty of Health Sciences at the University of Cape Town (UCT), South Africa. The research activities of Dr Georgia Schäfer's group primarily focus on oncogenic viruses relevant in the Sub-Saharan African context, an area that is additionally burdened by an HIV/AIDS epidemic of massive proportions. Her group has substantial research expertise on the early molecular mechanisms of Kaposi's Sarcoma Herpes Virus (KSHV) and Human Papillomavirus (HPV) infection, both viruses being associated with AIDS-defining malignancies, namely Kaposi's sarcoma and cervical cancer, respectively. Dr Georgia Schäfer's group is particularly interested in the molecules and receptors involved during early infection events that can potentially be developed into preventative and therapeutic targets. Her research involves both basic laboratory-based in vitro and in vivo studies (using recombinant KSHV virus or HPV virus-like particles, respectively, together with appropriate cell and/or animal models) as well as clinical studies involving nationally recruited patient cohorts. In response to the COVID-19 outbreak, Dr Schäfer's group has set up several SARS-CoV-2 diagnostic and pseudovirus neutralization assays to support various nation-wide research activities.

Communication

A Potential SARS-CoV-2 Variant of Interest (VOI) Harboring Mutation E484K in the Spike Protein Was Identified within Lineage B.1.1.33 Circulating in Brazil

Paola Cristina Resende [1,2,†], Tiago Gräf [1,3,*,†], Anna Carolina Dias Paixão [1,2], Luciana Appolinario [1,2], Renata Serrano Lopes [1,2], Ana Carolina da Fonseca Mendonça [1,2], Alice Sampaio Barreto da Rocha [1,2], Fernando Couto Motta [1,2], Lidio Gonçalves Lima Neto [4], Ricardo Khouri [1,3,5], Camila I. de Oliveira [1,3,5], Pedro Santos-Muccillo [1,3,6], João Felipe Bezerra [7], Dalane Loudal Florentino Teixeira [8], Irina Riediger [9], Maria do Carmo Debur [9], Rodrigo Ribeiro-Rodrigues [10,11], Anderson Brandao Leite [12], Cliomar Alves dos Santos [13], Tatiana Schäffer Gregianini [14], Sandra Bianchini Fernandes [15], André Felipe Leal Bernardes [16], Andrea Cony Cavalcanti [17], Fábio Miyajima [1,18], Claudio Sachhi [19], Tirza Mattos [20], Cristiano Fernandes da Costa [21], Edson Delatorre [1,22,‡], Gabriel L. Wallau [1,23,‡], Felipe G. Naveca [1,24,‡], Gonzalo Bello [1,25,‡] and Marilda Mendonça Siqueira [1,2,‡]

1 Fiocruz COVID-19 Genomic Surveillance Network, Fiocruz, Rio de Janeiro 21040-360, Brazil; paolabmrj@gmail.com (P.C.R.); carolinadiaspaixao@gmail.com (A.C.D.P.); luh.appolinario@gmail.com (L.A.); renata.serranolopes@gmail.com (R.S.L.); anacarolinafmend@gmail.com (A.C.d.F.M.); alicesampaio.br@gmail.com (A.S.B.d.R.); fcm@ioc.fiocruz.br (F.C.M.); ricardo_khouri@hotmail.com (R.K.); camila.indiani@fiocruz.br (C.I.d.O.); pedromuccillo@gmail.com (P.S.-M.); fabio.osv@gmail.com (F.M.); edson.delatorre@ufes.br (E.D.); gabriel.wallau@fiocruz.br (G.L.W.); naveca.felipe@gmail.com (F.G.N.); gbellobr@gmail.com (G.B.); mmsiq@ioc.fiocruz.br (M.M.S)

2 Laboratory of Respiratory Viruses and Measles (LVRS), Oswaldo Cruz Institute, Fiocruz, Rio de Janeiro 21045-900, Brazil

3 Plataforma de Vigilância Molecular, Instituto Gonçalo Moniz, Fiocruz-BA, Salvador 40296-710, Brazil

4 Laboratório Central de Saúde Pública do Estado do Maranhão (LACEN-MA), São Luís 65020-904, Brazil; lidio.neto@outlook.com

5 Laboratório de Enfermidades Infecciosas Transmitidas por Vetores, Instituto Gonçalo Moniz, Fiocruz-BA, Salvador 40296-710, Brazil

6 Laboratório de Patologia e Biologia Molecular, Instituto Gonçalo Moniz, Fiocruz-BA, Salvador 40296-710, Brazil

7 Laboratório de Vigilância Molecular Aplicada, Escola Técnica de Saúde, Centro de Ciências da Saúde, Universidade Federal da Paraíba, (UFPB), João Pessoa 58051-900, Brazil; jfb_rn@hotmail.com

8 Laboratório Central de Saúde Pública do Estado da Paraíba (LACEN-PB), João Pessoa 58013-360, Brazil; dalane.lacenpb@gmail.com

9 Laboratório Central de Saúde Pública do Estado do Paraná (LACEN-PR), São José dos Pinhais 83060-500, Brazil; irinariediger@sesa.pr.gov.br (I.R.); mariadebur@sesa.pr.gov.br (M.d.C.D.)

10 Laboratório Central de Saúde Pública do Estado do Espírito Santo (LACEN-ES), Vitoria 29050-755, Brazil; ro.ribeiro66@gmail.com

11 Núcleo de Doenças Infecciosas, Universidade Federal do Espírito Santo, Vitoria 29050-625, Brazil

12 Laboratório Central de Saúde Pública do Estado do Alagoas (LACEN-AL), Maceió 57036-860, Brazil; anderson.leite@icbs.ufal.br

13 Laboratório Central de Saúde Pública do Estado do Sergipe (LACEN-SE), Aracaju 49020-380, Brazil; cliomar.santos@fsph.se.gov.br

14 Laboratório Central de Saúde Pública do Estado do Rio Grande do Sul (LACEN-RS), Porto Alegre 90610-000, Brazil; tatiana-gregianini@saude.rs.gov.br

15 Laboratório Central de Saúde Pública do Estado de Santa Catarina (LACEN-SC), Florianópolis 88010-002, Brazil; sandrabiafe@gmail.com

16 Laboratório Central de Saúde Pública do Estado de Minas Gerais (LACEN-MG), Belo Horizonte 30510-010, Brazil; andre.leal@funed.mg.gov.br

17 Laboratório Central de Saúde Pública do Estado do Rio de Janeiro (LACEN-RJ), Rio de Janeiro 20231-092, Brazil; dg@lacen.fs.rj.gov.br

18 Oswaldo Cruz Foundation (Fiocruz), Branch Ceará, Eusebio 61760-000, Brazil

19 Instituto Adolfo Lutz (IAL), São Paulo 01246-000, Brazil; labestrategico@ial.sp.gov.br

20 Laboratório Central de Saúde Pública do Amazonas, Manaus 69020-040, Brazil; tirza_mattos@hotmail.com

21 Fundação de Vigilância em Saúde do Amazonas, Manaus 69060-000, Brazil; cristianozoo@yahoo.com.br

Citation: Resende, P.C.; Gräf, T.; Paixão, A.C.D.; Appolinario, L.; Lopes, R.S.; Mendonça, A.C.d.F.; da Rocha, A.S.B.; Motta, F.C.; Neto, L.G.L.; Khouri Cunha, R.; et al. A Potential SARS-CoV-2 Variant of Interest (VOI) Harboring Mutation E484K in the Spike Protein Was Identified within Lineage B.1.1.33 Circulating in Brazil. *Viruses* **2021**, *13*, 724. https://doi.org/10.3390/v13050724

Academic Editors: Burtram C. Fielding and Georgia Schäfer

Received: 18 March 2021
Accepted: 6 April 2021
Published: 21 April 2021

[22] Departamento de Biologia, Centro de Ciências Exatas, Naturais e da Saúde, Universidade Federal do Espírito Santo, Alegre 29500-000, Brazil
[23] Departamento de Entomologia e Núcleo de Bioinformática, Instituto Aggeu Magalhães (IAM), FIOCRUZ-PE-Recife 50740-465, Brazil
[24] Laboratório de Ecologia de Doenças Transmissíveis na Amazônia (EDTA), Leônidas e Maria Deane Institute, Fiocruz-AM, Manaus 69057-070, Brazil
[25] Laboratório de AIDS e Imunologia Molecular, Oswaldo Cruz Institute, Fiocruz, Rio de Janeiro 21040-900, Brazil
* Correspondence: tiago.graf@fiocruz.br
† These authors share the first authorship.
‡ These authors share the senior authorship.

Abstract: The severe acute respiratory syndrome coronavirus 2 (SARS-CoV-2) epidemic in Brazil was dominated by two lineages designated as B.1.1.28 and B.1.1.33. The two SARS-CoV-2 variants harboring mutations at the receptor-binding domain of the Spike (S) protein, designated as lineages P.1 and P.2, evolved from lineage B.1.1.28 and are rapidly spreading in Brazil. Lineage P.1 is considered a Variant of Concern (VOC) because of the presence of multiple mutations in the S protein (including K417T, E484K, N501Y), while lineage P.2 only harbors mutation S:E484K and is considered a Variant of Interest (VOI). On the other hand, epidemiologically relevant B.1.1.33 deriving lineages have not been described so far. Here we report the identification of a new SARS-CoV-2 VOI within lineage B.1.1.33 that also harbors mutation S:E484K and was detected in Brazil between November 2020 and February 2021. This VOI displayed four non-synonymous lineage-defining mutations (NSP3:A1711V, NSP6:F36L, S:E484K, and NS7b:E33A) and was designated as lineage N.9. The VOI N.9 probably emerged in August 2020 and has spread across different Brazilian states from the Southeast, South, North, and Northeast regions.

Keywords: SARS-CoV-2; E484K; variant of Interest; genomic epidemiology; Brazil

1. Introduction

The SARS-CoV-2 epidemic in Brazil was mainly driven by lineages B.1.1.28 and B.1.1.33 that probably emerged in February 2020 and were the most prevalent variants in most country regions until October 2020 [1,2]. Recent genomic studies, however, bring attention to the emergence of new SARS-CoV-2 variants in Brazil harboring mutations at the receptor-binding site (RBD) of the Spike (S) protein that might impact viral fitness and transmissibility.

So far, one variant of concern (VOC), designated as lineage P.1, and one variant of interest (VOI), designated as lineage P.2, have been identified in Brazil and both evolved from lineage B.1.1.28. The VOC P.1, first described in January 2021 [3], displayed an unusual number of lineage-defining mutations in the S protein (L18F, T20N, P26S, D138Y, R190S, K417T, E484K, N501Y, H655Y, T1027I) and its emergence was associated with a second COVID-19 epidemic wave in the Amazonas state [4,5]. The VOI P.2, first described in samples from October 2020 in the state of Rio de Janeiro, was distinguished by the presence of the S:E484K mutation in RBD and other four lineage-defining mutations outside the S protein [6]. The P.2 lineage has been detected as the most prevalent variant in several states across the country in late 2020 and early 2021 (https://www.genomahcov.fiocruz.br, accessed on 1 March 2021).

Several B.1.1.33-derived lineages are currently defined by the Pangolin system including: lineage N.1 detected in the US, lineage N.2 detected in Suriname and France, lineage N.3 circulating in Argentina, and lineages N.4 and B.1.1.314 circulating in Chile (https://cov-lineages.org/lineages.html, accessed on 1 March 2021). However, none of these B.1.1.33-derived lineages were characterized by mutations of concern in the S protein. Here, we define the lineage N.9 within B.1.1.33 diversity that harbors mutation E484K in the S protein as was detected in different Brazilian states between November 2020 and February 2021.

2. Materials and Methods

The Fiocruz COVID-19 Genomic Surveillance Network has recovered SARS-CoV-2 lineage B.1.1.33 genomes from 422 positive samples between 12th March 2020 and 27th January 2021 (Supplementary Material). Sequencing protocols were as previously described [7,8]. The FASTQ reads obtained were imported into the CLC Genomics Workbench version 20.0.4 (Qiagen A/S, Denmark), trimmed, and mapped against the reference sequence EPI_ISL_402124 available in EpiCoV database in the GISAID (https://www.gisaid.org/, accessed on 1 March 2021). The alignment was refined using the InDels and Structural Variants module.

Sequences were then combined with 816 B.1.1.33 Brazilian genomes available in the EpiCoV database in GISAID by 1st March 2021 (Supplementary Table S1). Only high quality (<1% of N) complete (>29 kb) SARS-CoV-2 genomes were used. This dataset was then aligned using MAFFT v7.475 [9] and subjected to maximum likelihood (ML) phylogenetic analysis using IQ-TREE v2.1.2 [10] under the GTR + F + G4 nucleotide substitution model, as selected by the ModelFinder application [11]. Branch support was assessed by the approximate likelihood-ratio test based on the Shimodaira–Hasegawa procedure (SH-aLRT) with 1000 replicates. The mutational profile was investigated using the Nextclade tool (https://clades.nextstrain.org, accessed on 1 March 2021) and temporal signal was assessed by the regression analysis of the root-to-tip genetic distance against sampling dates using the program Tempest [12].

A time-scaled phylogenetic tree was estimated using the Bayesian Markov Chain Monte Carlo (MCMC) approach implemented in BEAST 1.10.4 [13]. Bayesian tree was reconstructed using the GTR + F + I + G4 nucleotide substitution model, the non-parametric Bayesian skyline (BSKL) model as the coalescent tree prior and a strict molecular clock model with a uniform substitution rate prior (8×10^{-4}–10×10^{-4} substitutions/site/year). Ancestral node states were reconstructed using a reversible discrete phylogeographic model [14] where transitions between sampling locations (Brazilian states) were estimated in a continuous-time Markov chain (CTMC) rate reference prior. Convergence (effective sample size > 200) in parameter estimates was assessed using TRACER v1.7 18. The maximum clade credibility (MCC) tree was summarized with TreeAnnotator v1.10.4. ML and MCC trees were visualized using FigTree v1.4.4 (http://tree.bio.ed.ac.uk/software/figtree/, accessed on 1 March 2021).

3. Results and Discussion

Mutation profile analysis revealed a total of 34 B.1.1.33 sequences harboring the S:E484K mutation. ML phylogenetic analysis revealed that 32 of these sequences branched in a highly supported (SH-aLRT = 98%) monophyletic clade that define a potential new VOI designated as N.9 PANGO lineage [15]. The other two sequences harboring the S:E484K mutation branched separately in a highly supported (SH-aLRT = 100%) dyad (Figure 1a). The VOI N.9 is characterized by four non-synonymous lineage-defining mutations (NSP3:A1711V, NSP6:F36L, S:E484K, and NSP7b:E33A) and also contains a group of three B.1.1.33 sequences from the Amazonas state that has no sequencing coverage in the position 484 of the S protein, but share the remaining N.9 lineage-defining mutations (Table 1), thus forming a cluster of 35 sequences. The B.1.1.33 (S:E484K) dyad comprises two sequences from the Maranhao state and were characterized by a different set of non-synonymous mutations (Supplementary Table S2).

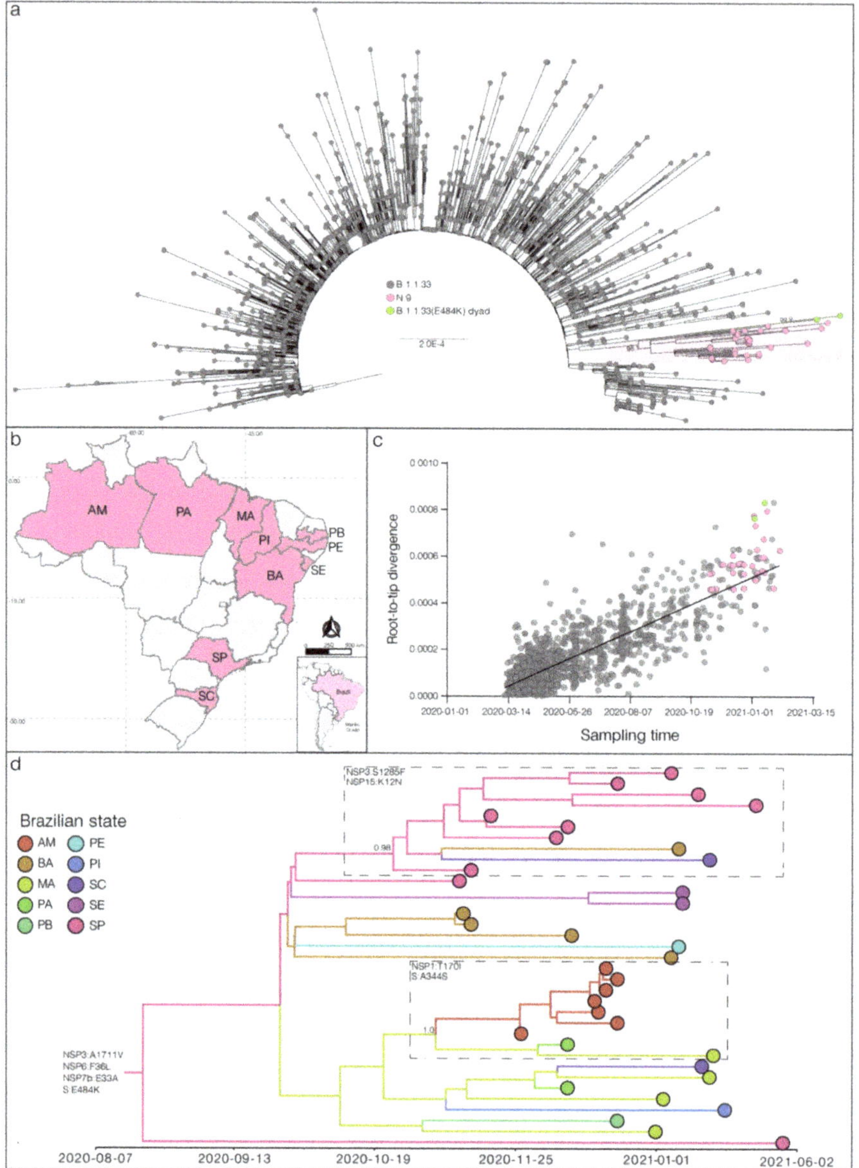

Figure 1. Lineage N.9 evolutionary origin and spatial-temporal distribution. (**a**) Maximum likelihood (ML) phylogenetic tree of the B.1.1.33 whole-genome sequences from Brazil. The B.1.1.33 sequences with mutation S:E484K are represented by pink (VOI N.9) and green (B.1.1.33(E484K)) circles. The SH-aLRT support values are indicated in key nodes and branch lengths are drawn to scale with the left bar indicating nucleotide substitutions per site. (**b**) Geographic distribution of the VOI N.9 identified in Brazil. Brazilian states' names follow the International Organization for Standardization (ISO) 3166-2 standard. (**c**) Correlation between the sampling date of B.1.1.33 sequences and their genetic distance from the root of the ML phylogenetic tree. Colors indicate the B.1.1.33 clade as indicated in (**a**). (**d**) Bayesian phylogeographic analysis of N.9 lineage. Tips and branches colors indicate the sampling state and the most probable inferred location of their descendent nodes, respectively, as indicated in the legend. Branch posterior probabilities are indicated in key nodes. Boxes highlight two N.9 subclades carrying additional mutations (indicated in each box). The tree was automatically rooted under the assumption of a strict molecular clock, and all horizontal branch lengths are time-scaled.

Table 1. Synapomorphic mutations of SARS-CoV-2 lineage N.9.

Genomic Region (Protein)	Nucleotide	Amino Acid
ORF1a	G1264T	-
ORF1a	C7600T	-
ORF1a (NSP3)	C7851T	A2529V (A1711V)
ORF1a (NSP6)	T11078C	F3605L (F36L)
Spike (S)	G23012A	E484K
ORF7b (NSP7b)	A27853C	E33A

Among the 35 genomes identified so far as VOI N.9, 10 Brazilian states were represented, suggesting that this lineage is already highly dispersed in the country. The VOI N.9 was first detected in Sao Paulo state on 11 November 2020, and soon later in other Brazilian states from the South (Santa Catarina), North (Amazonas and Para), and Northeast (Bahia, Maranhao, Paraiba, Pernambuco, Piaui, and Sergipe) regions (Figure 1b). Analysis of the temporal structure revealed that the overall divergence of lineage N.9 is consistent with the substitution pattern of other B.1.1.33 sequences (Figure 1c), thus suggesting no unusual accumulation of mutations in this VOI. Molecular clock analysis estimated the emergence of the VOI N.9 most probably in the states of Sao Paulo (Posterior State Probability (PSP) = 0.42), Bahia (PSP = 0.32) or Maranhao (PSP = 0.18) at 15th August, 2020 (95% High Posterior Density (HPD): 16th June–22th September, 2020) (Figure 1d). This analysis also revealed that some additional mutations were acquired during evolution of VOI N.9 in Brazil, determining two highly supported (PP > 0.95) subclades. One subclade, that mostly contains sequences from Sao Paulo state, probably arose on 16th October (95% HPD: 22th September–5th November) and was defined by additional mutations NSP3:S1285F and NSP15:K12N. The other subclade that mostly comprises sequences from the North region probably arose on 29th October (95% HPD: 5th October–17th November) and was defined by additional mutations NSP1:T170I and S:A344S (Figure 1d).

4. Conclusions

In this study we identified the emergence of a new VOI (S:E484K) within lineage B.1.1.33 circulating in Brazil. The VOI N.9 displayed a low prevalence (~3%) among all Brazilian SARS-CoV-2 samples analyzed between November 2020 and February 2021, but it is already widely dispersed in the country and comprises a high fraction (35%) of the B.1.1.33 sequences detected in that period. Mutation S:E484K has been identified as one of the most important substitutions that could contribute to immune evasion as confers resistance to several monoclonal antibodies and also reduces the neutralization potency of some polyclonal sera from convalescent and vaccinated individuals [16–18]. Mutation S:E484K has emerged independently in multiple VOCs (P.1, B.1.351 and B.1.1.7) and VOIs (P.2 and B.1.526) [19] spreading around the world, and it is probably an example of convergent evolution and ongoing adaptation of the virus to the human host.

The onset date of VOI N.9 here estimated around mid-August roughly coincides with the estimated timing of emergence of the VOI P.2 in late-July 6 and shortly precede the detection of a major global shift in the SARS-CoV-2 fitness landscape after October 2020 [20]. These findings indicate that 484K variants probably arose simultaneously in the two most prevalent viral lineages circulating in Brazil around July–August, but may have only acquired some fitness advantages, which accelerated its dissemination, after October 2020. We predict that the Brazilian COVID-19 epidemic during 2021 will be dominated by a complex array of B.1.1.28 (S:E484K), including P.1 and P.2, and B.1.1.33 (S:E484K) variants that will completely replace the parental 484E lineages that drove the epidemic in 2020. Implementation of efficient mitigation measures in Brazil is crucial to reduce community transmission and prevent the recurrent emergence of more transmissible variants that could further exacerbate the epidemic in the country.

Supplementary Materials: The following are available online at https://www.mdpi.com/article/10.3390/v13050724/s1, Supplementary Material: GISAID accession numbers of genomes lineage B.1.1.33 from this study, Supplementary Table S1: GISAID Acknowledgement Table, Supplementary Table S2: Synapomorphic non-synonymous mutations of the B.1.1.33(S:E484K) dyad isolated in the state of Maranhao.

Author Contributions: Conceptualization, P.C.R., T.G., E.D., G.L.W., F.G.N., G.B.; methodology, P.C.R., T.G., E.D., G.L.W., F.G.N., G.B.; formal analysis, P.C.R., T.G., E.D., G.L.W., F.G.N., G.B.; resources, F.G.N., M.M.S.; data acquisition and curation, A.C.D.P., L.A., R.S.L., A.C.d.F.M., A.S.B.d.R., F.C.M., L.G.L.N., R.K., C.I.d.O., P.S.-M., J.F.B., D.L.F.T., I.R., M.d.C.D., R.R.-R., A.B.L., C.A.d.S., T.S.G., S.B.F., A.F.L.B., A.C.C., F.M., C.S., T.M., C.F.d.C.; writing—original draft preparation, T.G., G.B.; writing—review and editing, P.C.R., T.G., E.D., G.L.W., F.G.N., G.B.; funding acquisition, F.G.N., M.M.S. All authors have read and agreed to the published version of the manuscript.

Funding: Financial support was provided by Fundação de Amparo à Pesquisa do Estado do Amazonas (FAPEAM) (PCTI-EmergeSaude/AM call 005/2020 and Rede Genômica de Vigilancia em Saúde-REGESAM); Conselho Nacional de Desenvolvimento Científico e Tecnológico (CNPq) (grant 402457/2020–0); CNPq/Ministério da Ciência, Tecnologia, Inovações e Comunicações/Ministério da Saúde (MS/FNDCT/SCTIE/Decit) (grant 403276/2020-9); Inova Fiocruz/Fundação Oswaldo Cruz (Grants VPPCB-007-FIO-18–2–30 and VPPCB-005-FIO-20–2–87), INCT-FCx (465259/2014–6) and Fundação Carlos Chagas Filho de Amparo à Pesquisa do Estado do Rio de Janeiro (FAPERJ) (26/210.196/2020). F.G.N, G.L.W, G.B and M.M.S are supported by the CNPq through their productivity research fellowships (306146/2017–7, 303902/2019–1, 302317/2017–1 and 313403/2018-0, respectively). G.B. is also funded by FAPERJ (Grant number E-26/202.896/2018).

Institutional Review Board Statement: The study was conducted according to the guidelines of the Declaration of Helsinki, and approved by the FIOCRUZ-IOC (68118417.6.0000.5248 and CAAE 32333120.4.0000.5190), the Amazonas State University Ethics Committee (CAAE: 25430719.6.0000.5016) and the Brazilian Ministry of the Environment (MMA) A1767C3.

Informed Consent Statement: A waiver of informed consent was obtained from the ethics committee.

Data Availability Statement: All genomes generated in this work were deposited in the EpiCoV database of GISAID (https://www.gisaid.org/, accessed on 1 March 2021). Accession codes are available in Supplementary Material.

Acknowledgments: The authors wish to thank all the health care workers and scientists who have worked hard to deal with this pandemic threat, the GISAID team, and all the EpiCoV database's submitters, GISAID acknowledgment table containing sequences used in this study is available in Supplementary Table S1. We also appreciate the support of the Fiocruz COVID-19 Genomic Surveillance Network (http://www.genomahcov.fiocruz.br/; accessed on 1 March 2021) members, the Respiratory Viruses Genomic Surveillance. General Coordination of the Laboratory Network (CGLab), Brazilian Ministry of Health (MoH), Brazilian States Central Laboratories (LACEN), Brazilian Ministry of Health (MoH), and the Amazonas surveillance teams for the partnership in the viral surveillance in Brazil.

Conflicts of Interest: The authors declare no conflict of interest.

Disclaimers: The opinions expressed by the authors do not necessarily reflect the opinions of the Ministry of Health of Brazil or the institutions with which the authors are affiliated.

References

1. Candido, D.S.; Claro, I.M.; de Jesus, J.G.; Souza, W.M.; Moreira, F.R.R.; Dellicour, S.; Mellan, T.A.; du Plessis, L.; Pereira, R.H.M.; Sales, F.C.S.; et al. Evolution and epidemic spread of SARS-CoV-2 in Brazil. *Science* **2020**, *369*, 1255–1260. [CrossRef] [PubMed]
2. Resende, P.C.; Delatorre, E.; Graf, T.; Mir, D.; Motta, F.C.; Appolinario, L.R.; Dias da Paixao, A.C.; da Fonseca Mendonca, A.C.; Ogrzewalska, O.; Caetano, B.; et al. Evolutionary dynamics and dissemination pattern of the SARS-CoV-2 lineage B.1.1.33 during the early pandemic phase in Brazil. *Front. Microbiol.* **2021**, *11*, 1–14. [CrossRef] [PubMed]
3. Fujino, T.; Nomoto, H.; Kutsuna, S.; Ujiie, M.; Suzuki, T.; Sato, R.; Fujimoto, T.; Kuroda, M.; Wakita, T.; Ohmagari, N. Novel SARS-CoV-2 variant identified in travelers from Brazil to Japan. *Emerg. Infect. Dis.* **2021**, *27*, 1243–1245. [CrossRef] [PubMed]

4. Naveca, F.; Nascimento, V.; Souza, V.; Corado, A.; Nascimiento, F.; Silva, G.; Costa, A.; Duarte, D.; Pessoa, K.; Mejia, M.; et al. COVID-19 epidemic in the Brazilian state of Amazonas was driven by long-term persistence of endemic SARS-CoV-2 lineages and the recent emergence of the new Variant of Concern, P.1. *Res. Sq* **2021**. preprint. Available online: https://www.researchsquare.com/article/rs-275494/v1 (accessed on 1 March 2021).
5. Faria, N.R.; Mellan, T.A.; Whittaker, C.; Claro, I.M.; da Silva Candido, D.; Mishra, S.; Crispim, M.A.E.; Sales, F.C.; Hawryluk, I.; McCrone, J.T.; et al. Genomics and epidemiology of a novel SARS-CoV-2 lineage in Manaus, Brazil. *medRxiv* **2021**. preprint. [CrossRef]
6. Voloch, C.M.; da Silva Francisco, R., Jr.; de Almeida, L.G.P.; Cardoso, C.C.; Brustolini, O.J.; Gerder, A.L.; Guimaraes, A.P.C.; Mariani, D.; da Costa, R.M.; Ferreisa, O.C., Jr.; et al. Genomic characterization of a novel SARS-CoV-2 lineage from Rio de Janeiro, Brazil. *J. Virol.* **2021**. [CrossRef] [PubMed]
7. Nascimento, V.A.; de Lima Guerra Corado, A.; do Nascimento, F.O.; da Costa, A.K.A.; Gomes Duarte, D.C.; Bessa Luz, S.L.; Goncalves, L.M.F.; de Jesus, M.S.; da Costa, C.F.; Delatorre, E.; et al. Genomic and phylogenetic characterization of an imported case of SARS-CoV-2 in Amazonas State, Brazil. *Mem. Inst. Oswaldo Cruz* **2020**, *115*, e200310. [CrossRef] [PubMed]
8. Resende, P.C.; Motta, F.C.; Roy, S.; Applinario, L.; Fabri, A.; Xavier, J.; Harris, K.; Matos, A.R.; Caetano, B.; Orgeswalska, M.; et al. SARS-CoV-2 genomes recovered by long amplicon tiling multiplex approach using nanopore sequencing and applicable to other sequencing platforms. *bioRxiv* **2020**. preprint. [CrossRef]
9. Katoh, K.; Standley, D.M. MAFFT multiple sequence alignment software version 7: Improvements in performance and usability. *Mol. Biol. Evol.* **2013**, *30*, 772–780. [CrossRef] [PubMed]
10. Minh, B.Q.; Schmidt, H.A.; Chernomor, O.; Schretmpf, D.; Woodhams, M.D.; von Haeseler, A.; Lanfear, R. IQ-TREE 2: New models and efficient methods for phylogenetic inference in the genomic era. *Mol. Biol. Evol.* **2020**, *37*, 1530–1534. [CrossRef]
11. Kalyaanamoorthy, S.; Minh, B.Q.; Wong, T.K.F.; Von Haetseler, A.; Jermiin, L.S. ModelFinder: Fast model selection for accurate phylogenetic estimates. *Nat. Methods* **2017**, *14*, 587–589. [CrossRef] [PubMed]
12. Rambaut, A.; Lam, T.T.; Carvalho, L.M.; Pybus, O.G. Exploring the temporal structure of heterochronous sequences using TempEst (formerly Path-O-Gen). *Virus Evol.* **2016**, *2*, vew007. [CrossRef] [PubMed]
13. Suchard, M.A.; Lemey, P.; Baele, G.; Ayres, D.L.; Drummond, A.J.; Rambaut, A. Bayesian phylogenetic and phylodynamic data integration using BEAST 1.10. *Virus Evol.* **2018**, *4*, vey016. [CrossRef] [PubMed]
14. Lemey, P.; Rambaut, A.; Drummond, A.J.; Suchard, M.A. Bayesian phylogeography finds its roots. *PLoS Comput. Biol.* **2009**, *5*, e1000520. [CrossRef] [PubMed]
15. Rambaut, A.; Holmes, E.C.; O'Toole, A.; Hill, V.; McCrone, J.T.; Ruis, C.; du Plessis, L.; Pybus, O.G. A dynamic nomenclature proposal for SARS-CoV-2 lineages to assist genomic epidemiology. *Nat. Microbiol.* **2020**, *5*, 1403–1407. [CrossRef] [PubMed]
16. Baum, A.; Fulton, B.O.; Wloga, E.; Copin, R.; Pascal, L.E.; Russo, V.; Giordano, S.; Lanza, K.; Negrom, N.; Ni, M.; et al. Antibody cocktail to SARS-CoV-2 spike protein prevents rapid mutational escape seen with individual antibodies. *Science* **2020**, *369*, 1014–1018. [CrossRef] [PubMed]
17. Greaney, A.J.; Loes, A.N.; Crawford, K.H.D.; Starr, T.N.; Malone, D.K.; Chu, H.Y.; Bloom, J.D. Comprehensive mapping of mutations in the SARS-CoV-2 receptor-binding domain that affect recognition by polyclonal human plasma antibodies. *Cell Host Microbe* **2021**, *29*, P463–P476.E6. [CrossRef] [PubMed]
18. Wang, P.; Liu, L.; Iketani, S.; Nair, M.S.; Luo, Y.; Guo, Y.; Wang, M.; Yu, J.; Zhang, B.; Kwong, P.D.; et al. Increased resistance of SARS-CoV-2 variants, B.1.351 and B.1.1.7 to antibody neutralization. *bioRxiv* **2021**. preprint. [CrossRef]
19. Gangavarapu, K.; Alkuzweny, M.; Cano, M.; Andersen, K.; Haag, E.; Hughes, L.; Mullen, J.; Su, A.; Latif, A.A.; Tsueng, G.; et al. outbreak.info. 2020. Available online: https://outbreak.info (accessed on 1 March 2021).
20. Martin, D.P.; Weaver, S.; Tegally, H.; San, E.L.; Shank, S.D.; Wilkinson, E.; Giandhari, J.; Naidoo, S.; Pullay, Y.; Singh, L.; et al. The emergence and ongoing convergent evolution of the N501Y lineages coincides with a major global shift in the SARS-CoV-2 selective landscape. *medRxiv* **2021**. preprint. [CrossRef]

MDPI

Article

SARS-CoV-2 Reverse Zoonoses to Pumas and Lions, South Africa

Katja Natalie Koeppel [1,2,†], Adriano Mendes [3,†], Amy Strydom [3,†], Lia Rotherham [4], Misheck Mulumba [4] and Marietjie Venter [3,*]

1 Department of Production Animal Studies, Faculty of Veterinary Science, University of Pretoria, Onderstepoort, Pretoria 0110, South Africa; katja.koeppel@up.ac.za
2 Centre for Veterinary Wildlife Studies, Faculty of Veterinary Sciences, University of Pretoria, Onderstepoort, Pretoria 0001, South Africa
3 Zoonotic Arbo- and Respiratory Virus Program, Centre for Viral Zoonoses, Department for Medical Virology, University of Pretoria, Pretoria 0110, South Africa; adriano.mendes288@gmail.com (A.M.); aimster.strydom@gmail.com (A.S.)
4 Onderstepoort Veterinary Research Institute, Agricultural Research Council, Onderstepoort, Pretoria 0110, South Africa; RotherhamL@arc.agric.za (L.R.); MulumbaM@arc.agric.za (M.M.)
* Correspondence: marietjie.venter@up.ac.za
† These authors contributed equally to this work.

Abstract: Reverse-zoonotic infections of severe acute respiratory syndrome-related coronavirus 2 (SARS-CoV-2) from humans to wildlife species internationally raise concern over the emergence of new variants in animals. A better understanding of the transmission dynamics and pathogenesis in susceptible species will mitigate the risk to humans and wildlife occurring in Africa. Here we report infection of an exotic puma (July 2020) and three African lions (July 2021) in the same private zoo in Johannesburg, South Africa. One Health genomic surveillance identified transmission of a Delta variant from a zookeeper to the three lions, similar to those circulating in humans in South Africa. One lion developed pneumonia while the other cases had mild infection. Both the puma and lions remained positive for SARS-CoV-2 RNA for up to 7 weeks.

Keywords: SARS-CoV-2; reverse zoonosis; wildlife

Citation: Koeppel, K.N.; Mendes, A.; Strydom, A.; Rotherham, L.; Mulumba, M.; Venter, M. SARS-CoV-2 Reverse Zoonoses to Pumas and Lions, South Africa. *Viruses* **2022**, *14*, 120. https://doi.org/10.3390/v14010120

Academic Editor: Ayato Takada

Received: 13 December 2021
Accepted: 2 January 2022
Published: 11 January 2022

Publisher's Note: MDPI stays neutral with regard to jurisdictional claims in published maps and institutional affiliations.

1. Introduction

Severe acute respiratory syndrome-related coronavirus 2 (SARS-CoV-2) is the causative agent of the disease COVID-19, which has caused a pandemic unlike any the present generation has seen before. SARS-CoV-2 belongs to the Coronaviridae family of positive sense single stranded RNA viruses [1]. The family consists of 46 species of virus, the majority of which have been isolated from animals [2]. Only seven viruses (NL63, 229E, HKU1, OC43, SARS, MERS and SARS-CoV-2) are known to infect humans but each are believed to have a zoonotic origin [3]. For this reason, as well as the high level of sequence similarity to a virus isolated for *Rhinolophus affinis* bats, RaTG13, it is believed that SARS-CoV-2 either descended directly from bats or evolved in a yet to be identified intermediate animal reservoir host before it was transmitted to humans [4,5]. The likely zoonotic spill-over highlights the importance of investigating transmission dynamics in animals to identify susceptible hosts but also define the risk for reverse zoonoses from humans and subsequent evolution.

Investigations of susceptibility of animals to SARS-CoV-2 can be categorised into three groups: those with predicted susceptibility, those experimentally infected, and naturally infected animals, with infections believed to occur through reverse zoonotic events. Studies which predict the susceptibility of animal species primarily utilise bioinformatics methods based on angiotensin converting enzyme 2 (ACE2) sequence homology, which is the receptor for the virus, and the SARS-CoV-2 Spike glycoprotein interaction [6–9]. Animals which have been successfully infected experimentally include domestic cats [10,11], ferrets [12,13], Rhesus macaques [14,15], fruit bats [13], golden Syrian hamsters [16,17] and deer mice [18]. Animal surveillance programs have also discovered reverse zoonotic events in domestic species including cats [19,20] and dogs [19], as well as captive wildlife populations such as mink [21], otter [22], ferrets [23], lions [24,25], tigers [24], snow leopards, gorillas [22,26], and white-tailed deer [27]. With the emergence of SARS-CoV-2 variants of concern, the question as to whether evolution of the virus will favour reverse-zoonoses and animal transmission is important to address.

South Africa experienced three waves of infection of COVID-19 from March 2020 to October 2021. Wave 1 was characterised by a mixture of the original strains, wave 2 by the beta variant and wave 3 by the Delta variant [28,29]. South Africa has a lucrative wildlife industry mainly based on conservation and tourism with several large feline species being kept in wildlife reserves but also in zoos across the country. Here we describe the natural infection of SARS-CoV-2 in a puma during the first wave and three lions during the third wave in a private zoo in South Africa in at least three transmission events from their handlers. The three lions were all infected with the Delta variant while the puma was infected during the first wave but not genetically investigated. With lions and other big cats being found naturally in wildlife reserves as well as higher density settings in South Africa, the risk to these animals being infected from humans either through close contact or through handling of food requires further attention. It is also equally important to assess the risk of subsequent transmission between animals and prolonged shedding that may give rise to new variants.

2. Materials and Methods

2.1. Ethics Statement

The study was approved by the Human and Animal Ethics Committee of the University of Pretoria (REC150-20) and Section 20 application by the Department of Agriculture Land Reform and Rural development of South Africa (12/11/1/1/8 (1612 AC)).

2.2. Outbreak Description

In July 2020, two pumas (LPZ0017 and LPZ0018) in a private zoo showed signs of anorexia, diarrhoea, and nasal discharges. The two pumas were in one enclosure. LPZ0018 also developed ocular discharge and a dry cough, which persisted for 13 days. LPZ0018 was anaesthetised with medetomidine (2 mg, Kyron Laboratories, Johannesburg, South Africa) and zolazepam and tiletamine (Zoletil®, Virbac, South Africa, 40 mg) on 27 July 2020. A nasopharyngeal sample (NP) was taken for SARS-CoV-2 after 5 days of persistent coughing, which did not respond to antibiotic therapy (12 mg/kg bid, Amoxycillin/Clavulanic acid, Auro Amoxiclav, Aurobindo, Johannesburg, South Africa). Follow-up samples were taken on the 13 August, 25 August, and 9 September 2020 of LPZ0018. As puma LPZ0017 only presented with mild clinical signs, it was decided not to anaesthetise the animal for testing purposes. Both pumas made a full recovery after 23 days.

In June 2021, three lions (ZRU125/21, ZRU127/21 and ZRU128/21) who were all born in captivity and raised in a zoo exhibited respiratory symptoms. ZRU127/21 and ZRU128/21 were kept in one enclosure and ZRU125/21 was kept in a separate enclosure. The clinical signs in these lions were predominantly upper respiratory with nasal and ocular discharge and a dry cough for up to 14 to 15 days. A persistent cough was seen between 5 and 15 days with worsening and difficulty breathing in two lions (ZRU127/21 and ZRU128/21) for 10 days after the onset of cough. Transient anorexia (1 to 2 days) was seen in 2 out of 3 lions (ZRU125/21 and ZRU127/21). Lions were treated orally with amoxiclav (8 mg/kg bid) and a NSAID (meloxicam, 0.05 mg/kg, qd, coxflam, Novartis, Johannesburg, South Africa). ZRU125/21 did not respond to antibiotics (Amoxycillin/Clavulanic acid, Auro Amoxiclav, Aurobindo, Johannesburg, South Africa). The lioness was immobilized with medetomidine (6 mg) and zolazepam and tiletamine (Zoletil®, 100 mg) and a NP sample was tested for SARS-CoV-2 on 22 June. Subsequent oropharyngeal or NP samples were taken on 25 June of the other two lions as well as zoo staff who had direct or indirect contact with the lions (Supplementary Table S1). Staff and lions were monitored in the subsequent weeks for the presence of SARS-CoV-2. Voided faecal samples from the lions were also collected from 25 June to 12 July 2021 (Supplementary Table S2). ZRU125/21 received a dose of dexamethasone (Kortico, Bayer, Johannesburg, South Africa) intravenously as she started to develop pneumonia indicated by bronchial changes on radiographs (Figure 1B). All three lions made a full recovery within 15 to 25 days.

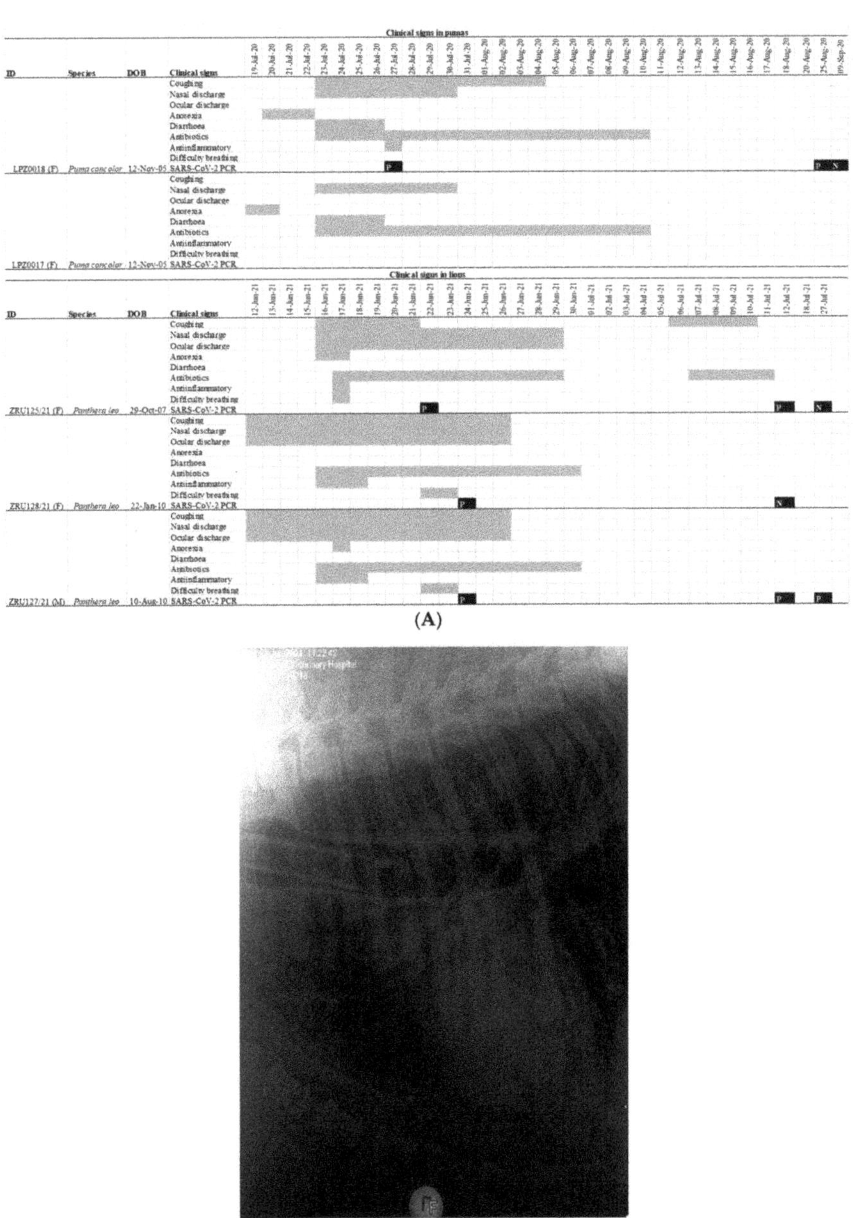

Figure 1. (A): Clinical features and timeline of SARS-CoV-2 infection in the pumas and lions. Grey bars

indicate the duration of a sign of infection. Black squares indicate the date of a RT-PCR test with a P indicating a positive test and N a negative one. All RT-PCR results shown are for nasopharyngeal swabs. (**B**): Lateral view of the chest of the 14-year-old lion ZRU125/21 showing marked bronchial lung pattern suggestive of bronchopneumonia.

2.3. RT-PCR of SARS-CoV-2

Dry nasal swabs were placed into 1 mL of PBS and left at room temperature for 10–30 min. Thereafter the samples were vortexed for 1 min. Nucleic acid was extracted from NP or faecal samples with the Qiamp Viral Mini Kit (Qiagen, Hilden, Germany) according to the manufacturer's instructions.

The AllpexTM 2019-nCoV assay was used to test for the presence of SARS-CoV-2 in the puma samples at the Agricultural Research Council, Onderstepoort Veterinary Research Institute. Briefly, 8 uL of extract was mixed with 5 uL reaction buffer, 5 uL water, 5 mL of primer and probe mixture, and 2 uL of enzyme mix. The samples were run on a Biorad Cfx96 (Biorad, Hercules, CA, USA) or a Rotor-Gene Q 5plex Platform (Qiagen, Montgomery, MD, USA). Results were analysed using CFX Manager Software (Biorad, Hercules, CA, USA) or Q-Rex Software (Qiagen, Montgomery, MD, USA).

LightMix SarbecoV E-gene and RdRp gene kits (TIP-MOLBIOL, Berlin, Germany) were used to test for the presence of SARS-CoV-2 in lion samples at the Zoonotic, Arbo- and Respiratory Virus research program, Centre for Viral Zoonoses, University of Pretoria. Briefly, 10 uL of template was mixed with 0.5 uL parameter specific reagent, 0.8 uL AgPath-ID™ One-Step RT-PCR Reagents (Applied Biosystems™, Waltham, MA, USA), and 10 uL buffer. Reactions were run on a ViiA7 (Thermo Fisher Scientific, Waltham, MA, USA). Results were analysed using QuantStudio Real-Time PCR Software (Thermo Fisher Scientific, Waltham, MA, USA). The AllpexTM 2019-nCoV assay and LightMix SarbecoV E-gene and RdRp gene kits are approved by the South African Health Products Regulatory Authority (SAHPRA).

2.4. Whole Genome Sequencing and Phylogenetic Analysis

Nucleic acid was extracted from 200 uL NP samples with the Qiamp Viral Mini Kit (Qiagen) according to the manufacturer's instructions in a total elution volume of 50 uL. The RNA clean and concentrator 5 kit (Zymo Research, Irvine, CA, USA) was used to concentrate RNA to a final volume of 15 uL. Ct values were determined with the LightMix SarbecoV E-gene kit using 3 uL RNA and 7 uL nuclease free water. Complementary DNA was synthesized with Superscript IV Reverse Transcriptase (Thermo Fisher Scientific, Waltham, MA, USA) and random primers according to the manufacturer's instructions. The SARS-CoV-2 genomes were amplified with a 1:1 combination of the Artic primer pools V2 and V3 (https://artic.network/ncov-2019 (accessed on 14 July 2021)). The reaction was done with Q5 High fidelity polymerase (New England Biolabs, Ipswich, MA, USA). PCR clean-up was done with Agencourt AMPure XP beads (Beckman Coulter, Carlsbad, CA, USA) according to the manufacturer's instructions. Libraries were prepared with the Illumina DNA prep kit (Illumina, Berlin, Germany) using between 100 and 500 ng of cDNA. Tagmentation was done at 55 °C for 15 min and amplification of tagmented DNA was done with the enhanced PCR mix and index adapters. Libraries were purified and size selected with the sample purification beads and 80% ethanol. Libraries were normalized to a final concentration of 12 pM before sequencing using an iSeq 100 i1 Reagent v2 (300 cycles) kit.

FASTQ files were uploaded to the Galaxy web platform and the public server at usegalaxy.eu to analyse the data [30] The Galaxy workflow for the analysis of Illumina paired end sequenced ARTIC amplicon data was used to assemble raw data [30].

2.5. Sanger Sequencing of the Spike Gene

cDNA was synthesized from RNA with Superscript III Reverse Transcriptase (Invitrogen™, Waltham, MA, USA) and 5 uM of SARS2_S R1 (GGAGACACTCCATAACACTTAA).

First round amplification was done with 10 uM of SARS2_S_F1 (GTCTCTAGTCAGTGT-GTTA) and 10 uM SARS2_S_R1 and the Platinum™ II Taq Hot-Start DNA Polymerase kit (Invitrogen™, Waltham, MA, USA). Then, 5 uL of cDNA was mixed with 4 uL 10x buffer, 0.4 uL of dNTP as well as each primer, 0.16 uL Taq, and 9.64 molecular grade water. Thermal cycling parameters were as follows: initial denaturation at 94 °C for 2 min; 35 cycles of denaturation at 94 °C for 15 s, annealing at 50 °C for 15 s, and elongation at 68 °C for 15 s. The product (2 uL) was used as template for a second-round of amplification with 10 uM of each primer: SARS2_S_F2 (CCCCTGCATACACTAATTCTT) and SARS2_S_R2 (AAACTTCACCAAAAGGGCACAAG). The reaction volumes were the same as in the first round with a total reaction volume of 20 uL. Conditions were also the same except annealing at 55 °C and resulted in a 900 kb product. PCR products were purified and sequenced at Inqaba Biotec, South Africa.

2.6. Phylogenetic Analyses

The 'Phylogenetic Assignment of Named Global Outbreak Lineages' (PANGOLIN) software suite (https://github.com/hCoV-2019/pangolin (accessed on 21 July 2021)) was used for SARS-CoV-2 lineage classification [31]. Outbreak strains were compared to global Delta sequences collected between May and June 2021. The 43,322 sequences were subsampled to 1545 in Nextstrain based on genetic proximity to the study sequences [32]. Sequences were aligned in ViralMSA and a maximum likelihood tree was inferred with IQ-TREE and the GTR + G4 model [33,34]. To determine the time to the most recent common ancestors of the study strains, a maximum likelihood (IQ-TREE) tree was inferred from 115 South African Delta sequences. TempEst v1.5.3 was used to plot the root-to-tip genetic distance and sampling dates and sequences that did not conform to a linear evolutionary pattern were removed from the dataset [35]. Bayesian phylogenetic inference was done in BEAST v1.10.5 using the GTR + G4 substitution model under a relaxed clock and coalescent Gaussian Markov random field (GMRF) [36]. A 100 million steps for a Markov chain Monte Carlo chain was run and every 10,000th generation was sampled. This was repeated twice. A maximum clade credibility (MCC) tree was summarized using TreeAnnotator v1.10.5. Trees were visualized and annotated using the FigTree (v1.4) program and Microreact [37].

3. Results

3.1. Clinical Features of SARS-CoV-2 Infection in Pumas and Lions

On 19 July 2020 one 12-year-old female puma (LPZ0018) in a private zoo in Johannesburg, Gauteng province of South Africa, showed signs of anorexia, followed 24 h later by a second puma (LPZ0017) with similar signs (Figure 1A). Nearly a year later, on the 12 June 2021, two lions (ZRU127/21 and ZRU128/21) became sick with nasal and ocular discharge and coughing in the same private zoo. Four days later, a 14-year-old female lion (ZRU125/21) housed in a separate enclosure also became sick (Figure 1A). This lioness developed lower respiratory tract infection with signs of bronchial pneumonia (Figure 1B). These animals were unresponsive to antibiotic therapy. NP samples from one puma (LPZ0018) and all three lions tested positive for SARS-CoV-2 by real time reverse transcriptase polymerase chain reaction (RT-PCR). Upon diagnosis, LPZ0018 was treated with doxycycline intramuscular (Bayer Animal Health, South Africa), a non-steroidal anti-inflammatory drug (meloxicam, 0.05 mg/km subcutaneously, Boehringer Ingelheim, South Africa) and a vitamin supplement (0.1 mL/kg, Kyroligo, Kyron Laboratories, South Africa). This resulted in an improvement in condition despite NP samples remaining positive for 4 weeks on faecal swabs and 6 weeks on nasal swabs. The two lions, ZRU127/21 and ZRU128/21, were initially also placed on oral antibiotics and non-steroidal anti-inflammatory drugs (Meloxicam) (Figure 1A). ZRU125/21 was treated with a single dose of dexamethasone (Kortico, Bayer Ltd., Johannesburg, South Africa) and antibiotics (Draxxin, Tulrathromycin, 200 mg SC, Zoetis, South Africa) for secondary bacterial infection. All three lions fully recovered approximately 3 weeks post disease onset. The lions were placed under quarantine from the time they tested positive until they were cleared from infection (7 weeks post onset of signs).

3.2. Kinetics of SARS-CoV-2 Infection

The puma and the lions were monitored for 10 weeks (puma) and 5 weeks (lions) after initial identification of SARS-CoV-2 infection. Both NP (Figure 2) and faecal samples (Supplementary Table S1) were used for RT-PCR testing. Except for the RdRp target (Ct: 23.2), the Ct values for the puma (LPZ0018) infection were already high (E gene Ct: 34 and N gene Ct: 36) when the first sample was taken, suggesting that the animal was towards the end of acute infection. Ct values in the mid-30s across all targets were still present after 1 month (6 weeks after initial signs) (Figure 2A). The virus was undetectable by 9 September 2020. Due to the low level of viral RNA, we were unable to use this sample for whole genome sequencing. The second puma (LPZ0017) only had mild clinical signs and no cough and thus it was decided not to anaesthetise the animal at that time for testing.

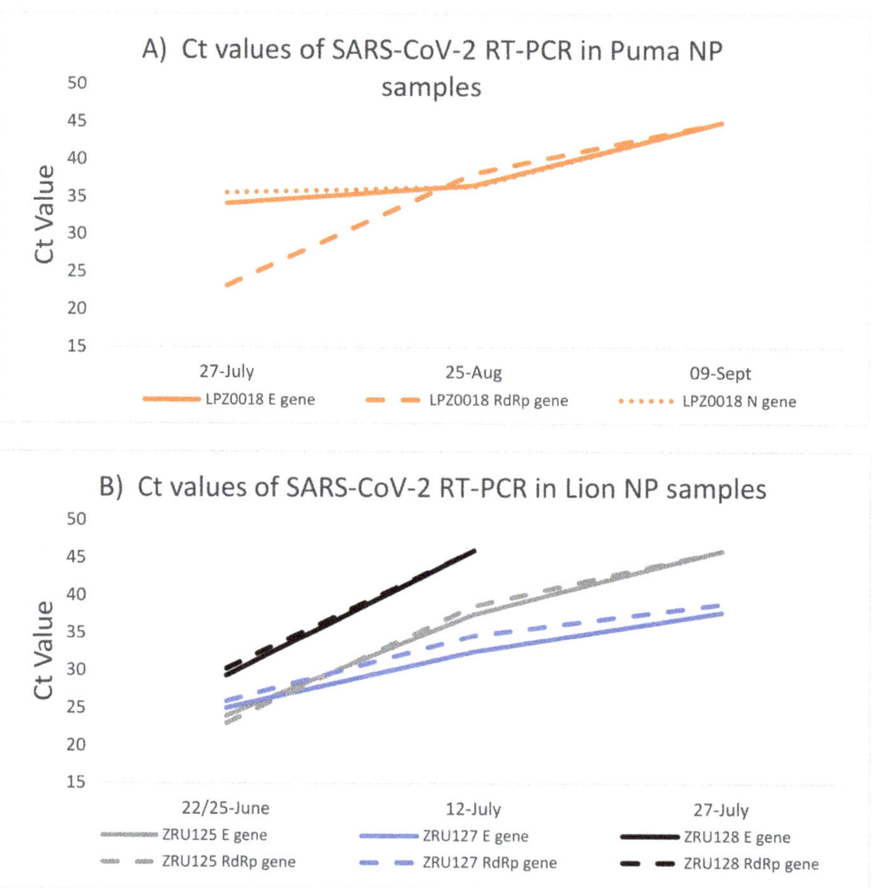

Figure 2. SARS-CoV-2 RNA load in puma and lions measured by RT-PCR over time. (**A**): Ct values detected in NP samples of the puma (LPZ0018). (**B**): Ct values detected in NP samples of the three lions. NP—Nasopharyngeal swab; E—SARS-CoV-2 envelope gene (solid lines); RdRp -SARS-CoV-2 RNA dependent RNA polymerase gene (dashed lines); N—SARS-CoV-2 nucleocapsid gene (dotted lines).

Two of the lions, ZRU125/21 and ZRU127/21, exhibited similar infection kinetics with a peak of viral RNA load detected (E gene Ct: 24 and 25) (Figure 2B) at the first collection point. ZRU128/21 showed lower levels of viral RNA at this same time point (Ct: 29.3). Less RNA was detected as the lions recovered from disease; however, viral RNA could still be detected in the nasopharynx of ZRU127/21 5 weeks (27 July 2021) after the first sample was taken (Ct: 37.84) i.e., 7 weeks after onset of clinical signs.

Lion faecal samples tested positive for 3 weeks after clinical signs appeared. No faecal samples tested PCR positive after 7 July 2021. This suggested that the lions were clear of viral RNA in their faeces approximately 1 month after first signs (Supplementary Table S2). No viral RNA was detected in the blood. In combination, these data show that SARS-CoV-2 infected and replicated in the respiratory and gastrointestinal tract in at least one puma and all three lions, concomitant with symptomology that is similar to COVID-19 in humans.

3.3. Investigation of Human–Lion Transmission Route

A One Health investigation into the source of infection to the lions was conducted on twelve members of staff who had either been in direct or indirect contact with the lions through structured interviews and collection of respiratory samples following informed consent. Both RT-PCR (NP/OP swabs) and ELISA (serum) testing were carried out (Supplementary Table S1). One staff member with direct contact (ZRUCWL005) and one with indirect contact (ZRUCWL012) tested PCR positive for SARS-CoV-2 on 25–26 June 2021 (approximately 2 weeks after the start of the lion disease course and while all three lions were PCR positive). These two individuals and three more staff members (a total of five staff members) also tested positive for anti-Spike IgG antibodies. None of the staff interviewed reported any recent symptoms of COVID-19. Follow-up samples were collected from the two PCR positive members of staff 17 days after the first tests. Both follow-up samples were still positive by PCR with Ct values of 33.30 (ZRUCWL005) and 35.95 (ZRUCWL012). These data suggest that SARS-CoV-2 was circulating amongst the staff during the time at which the lions got sick and suggests that staff members with direct contact with the lions were likely responsible for transmission.

3.4. Genomics of SARS-CoV-2 in the Lion Outbreak

In order to determine if the staff and lions were infected with the same strain and to shed light on the route of transmission, genome sequencing was conducted on both humans and the three lions. We obtained near full-length sequences (92.3–98.4%) for all five samples with gaps in the Spike gene filled in by sanger sequencing. All five sequences had between 99.93 and 100% nucleotide identity. NextClade analysis as well as multiple sequence alignment (MSA) of the Spike glycoproteins revealed that each of the infections was a Delta variant (B.1.617.1) of SARS-CoV-2. More specifically, the three lions and ZRUCWL005 were classified as the AY.38 lineage of Delta (Figure 3). Phylogenetic analysis comparing the study sequences with local and global strains confirmed that all three lions and the keeper, ZRUCWL005, clustered together with South African sequences while ZRUCWL012 clustered in a separate South African clade (Figure 3). Additionally, the SARS-CoV-2 sequences detected in the South African lions were divergent from the Delta sequences detected in India from an outbreak in April/May 2021 [25]. Bayesian analysis indicated that the time of the Most Recent Common Ancestor (tMRCA) between the lions and ZRUCWL005 was around the end of May (95% HPD 2021.35-2021.45) and clustered with sequences detected in Gauteng, KwaZulu Natal, and Limpopo provinces of South Africa (Supplementary Figure S1). ZRUCWL012 shared a MRCA with the rest of the study strains around the middle of April (95% HPD 2021.15-2021.32).

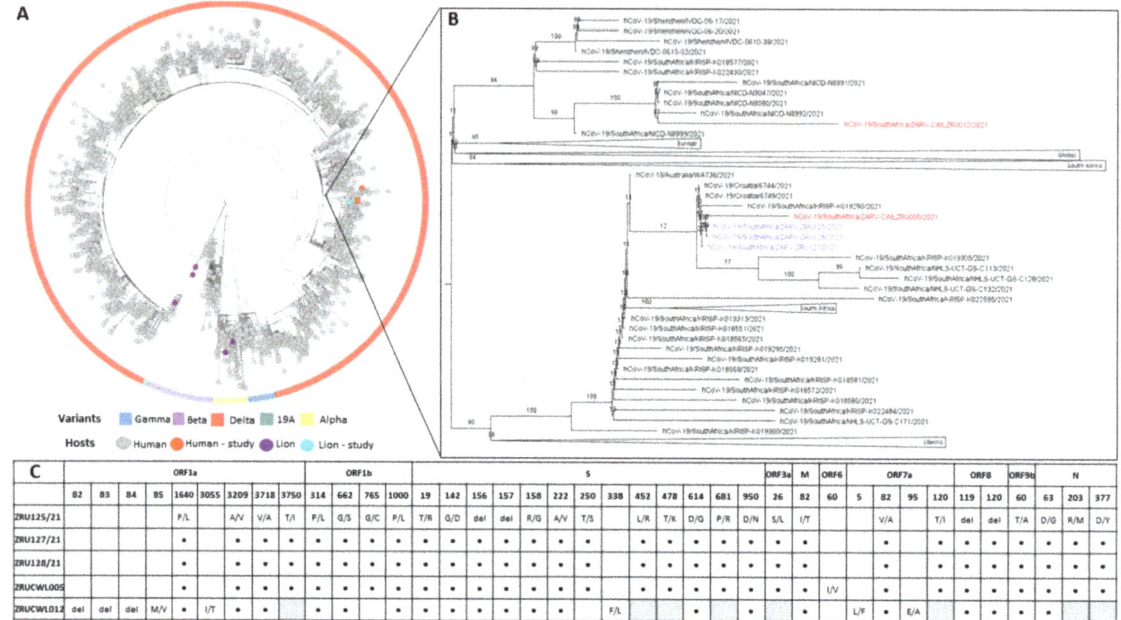

	ORF1a									ORF1b				S													ORF3a	M	ORF6	ORF7a				ORF8		ORF9b	N		
	82	83	84	85	1640	3055	3209	3718	3750	314	662	765	1000	19	142	156	157	158	222	250	338	452	478	614	681	950	26	82	60	5	82	95	120	119	120	60	63	203	377
ZRU125/21					P/L		A/V	V/A	T/I	P/L	G/S	G/C	P/L	T/R	G/D	del	del	R/G	A/V	T/S		L/R	T/K	D/G	P/R	D/N	S/L	I/T			V/A		T/I	del	del	T/A	D/G	R/M	D/Y
ZRU127/21					•		•	•	•	•	•	•	•	•	•	•	•	•	•	•		•	•	•	•	•	•	•			•		•	•	•	•	•	•	•
ZRU128/21					•		•	•	•	•	•	•	•	•	•	•	•	•	•	•		•	•	•	•	•	•	•			•		•	•	•	•	•	•	•
ZRUCWL005					•		•	•	•	•	•	•	•	•	•	•	•	•	•	•		•	•	•	•	•	•	•	I/V		•		•	•	•	•	•	•	•
ZRUCWL012	del	del	del	M/V	•	I/T	•	•		•	•		•	•	•	•	•	•	•		F/L			•		•		•		L/F	•	E/A		•	•	•	•		

Figure 3. (**A**): Maximum likelihood tree of global SARS-CoV-2 genomes. Variants of concern are indicated in the outer circle and host species are indicated by coloured tips. (**B**): View of branch where SARS-CoV-2 sequences determined in this study clustered. Sequences from lions are in blue and sequences from humans are in red: ZRU125/21 (EPI-ISL-6261983), ZRU127/21 (EPI-ISL-6261987), ZRU128/21 (EPI-ISL-6261989), ZRUCWL005 (EPI-ISL-6261993) and ZRUCWL012 (EPI-ISL-6261996). (**C**): Amino acid changes in study strains when compared to the Wuhan-Hu-1 reference genome (NC_045512.2). Black dots represent identical changes and grey boxes represent missing data. Sequencing was done from NP swabs.

4. Discussion

The results presented in this study document outbreaks of SARS-CoV-2 in pumas and lions kept in a South African private zoo. The two pumas and three lions presented with respiratory illness which was similar to COVID-19 in humans. Clinical signs in the animals in this report ranged from mild influenza-like illness including cough, to difficulty breathing and pneumonia. Additionally, both pumas and the three lions presented early on with ocular and/or nasal discharge, a sign that may be distinctive from human infection. The animals did not respond to antibiotic treatment but made uneventful recoveries following treatment with anti-inflammatory drugs and supportive care. Cases reported in large felids in this study are considered mild and are similar to mild infections in 32 confirmed positive large felids housed in zoological collections from April 2020 to August 2021 [38].

Detection of viral RNA in both the upper respiratory tract and the faeces as well as the fact that the pumas and the lions presented with the concomitant symptoms illustrates that this virus is able to generate a bona fide infection within these animals via a natural infection route. Despite extended viral shedding, all of the infected cats recovered fully. These outbreaks are at least the third and fourth of its kind in which SARS-CoV-2 has been shown to transmit between humans and captive large felines, although the current study is the only to report on genomic One Health investigations of Delta variants transmitted from humans to animals. These reports, as well as the evidence of experimental infection, make it clear that large felids are particularly susceptible to this virus.

Unfortunately, we were unable to carry out an investigation into the source or the specific variant involved in the puma outbreak. The samples were diagnosed by real-time PCR at the time of the outbreak, but when we attempted to sequence the sample a year later there was insufficient RNA left for genome sequencing. A One Health epidemiological investigation on the lion outbreak indicated that two staff members of the zoo also had SARS-CoV-2 in 2021. Whole genome sequencing and phylogenetic analysis indicated that ZRUCWL005, detected in a staff member who had direct contact with the lions, was closely related to the lion sequences. The genome sequence of the second staff member (ZRUCWL012) who did not have direct contact with the lions was slightly divergent and seems not to be part of the reverse zoonotic outbreak. There were little to no differences in nucleotide identity (99.93–100%) between the lion sequences and ZRUCWL005. This indicates that unlike in mink and in this case, a host switch did not result in evolutionary pressure to change the sequence of the SARS-CoV-2 variant [21] although only early sequences were obtained from the lions.

The timeline of infections for lions and ZRUCWL005 is difficult to estimate since all staff members were asymptomatic during the outbreak. RT-PCR results from the study might indicate that the lion and human infections occurred more or less in parallel. This leads us to believe that the transmission route for this outbreak is either ZRUCWL005 (the head big cat keeper) or another human contact, although the true direction of transmission is difficult to estimate. Three other staff members tested positive for anti-Spike IgG antibodies and only one reported a previous known positive COVID-19 test. This result was in January 2021, which was too long before the outbreak to be related. It is therefore also possible that one of these staff members infected ZRUCWL005 and the lions concomitantly (Figure 4). The index infection was likely in May when these sequences shared a MRCA. Since ZRU127/21 and ZRU128/21 presented with clinical signs on the same day, it is likely that each was infected by the original source. It is also possible that the original index case, whether identified or not in this study, transmitted the virus to the lions which subsequently passed it on to ZRUCWL005. Isolation of infectious virus from lion swabs was inconclusive and it was not possible to determine whether the lions were shedding infectious virus at the time of sampling. It is, however, clear that at least two reverse zoonotic events occurred in June 2021 in this zoo since ZRU125/21 was kept in a separate cage with no contact to the other two lions.

Reverse zoonotic transmission of SARS-CoV-2 from asymptomatic animal handlers pose a risk to large felines kept in captivity. Transmission of the Delta variant to these animals may potentially result in more severe disease. Prolonged shedding may spread the virus to animals in close proximity. Precautions should be implemented in zoos and other settings where these animals may have frequent exposures to humans to prevent such events and in particular to avoid introduction of SARS-CoV-2 to the wider population of animals in the wild where control measures are difficult to implement sufficiently early.

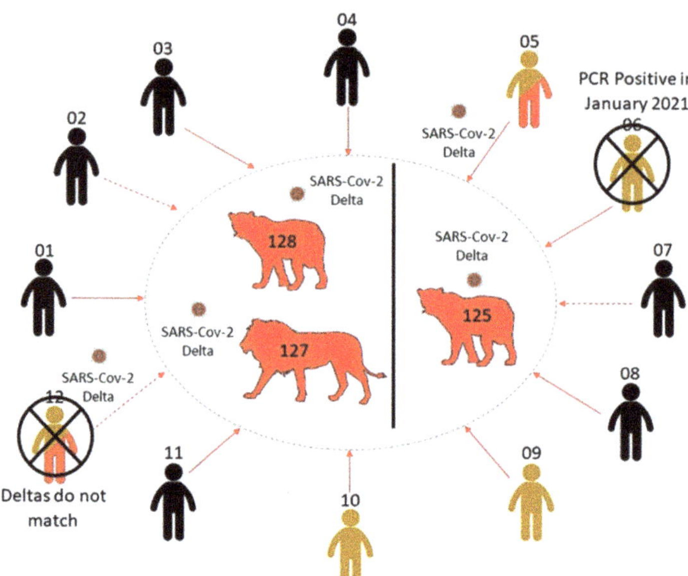

Figure 4. A summary of the potential infection route from animal handlers to the three lions. Direct (solid line) and indirect (dashed line) human contacts were traced and tested for SARS-CoV-2 RNA and IgG antibodies. Negative cases are coloured in black. PCR positive cases are coloured in red, while serologically positive cases are coloured in yellow. The two contact cases which were unlikely to be responsible for infecting the lions, owing to differing Delta sequences (ZRUCWL012) and previous positive tests (ZRUCWL006) are marked with a cross.

Supplementary Materials: The following are available online at https://www.mdpi.com/article/10.3390/v14010120/s1, Table S1: PCR and serological results from direct and indirect human contacts with infected lions, Table S2: Detection of SARS-CoV-2 E and RdRp genes from lion faecal samples in Ct values, Figure S1: Bayesian phylogenetic inference of whole genome sequences detected in lions and humans.

Author Contributions: Conceptualization, K.N.K. and M.V.; methodology, A.M., A.S., K.N.K. and L.R.; software, A.S. and A.M.; validation, A.S. and A.M.; formal analysis, A.S. and A.M.; investigation, A.M., A.S., K.N.K., M.V. and L.R.; resources, M.V. and K.N.K.; data curation, A.S. and A.M.; writing—original draft preparation, A.M., A.S., K.N.K. and M.V.; writing—review and editing, A.M., A.S., K.N.K., M.V., L.R. and M.M.; visualization, A.S., A.M. and L.R.; supervision, M.V.; project administration, A.M., A.S., K.N.K. and M.V.; funding acquisition, M.V. All authors have read and agreed to the published version of the manuscript.

Funding: This study was funded through the G7 Global Health fund supplementary funding for SARS-CoV-2 through the Robert Koch Institute for the African Network for improved diagnostics and epidemiology of common and emerging infections (ANDEMIA, Grant: G7 (TRICE): ZMVI1-2517GHP703).

Institutional Review Board Statement: The study was approved by the Human and Animal Ethics Committee of the University of Pretoria (REC150-20) and Section 20 application by the Department of Agriculture Land Reform and Rural development of South Africa (12/11/1/1/8 (1612 AC)).

Informed Consent Statement: Informed consent was obtained from all subjects involved in the study.

Data Availability Statement: Whole genome sequences were uploaded to GISAID with the following accession numbers: ZRU125/21 (EPI-ISL-6261983), ZRU127/21 (EPI-ISL-6261987), ZRU128/21 (EPI-ISL-6261989), ZRUCWL005 (EPI-ISL-6261993) and ZRUCWL012 (EPI-ISL-6261996).

Acknowledgments: We would like to thank the zoo staff for allowing us to investigate this outbreak as well as veterinarians and animal technicians that supported us. We also thank Kreshalen Govender and Frank Fox for their support in specimen collection and M Quan for running tests at Department of Veterinary Tropical Disease.

Conflicts of Interest: The authors declare no conflict of interest.

References

1. Walker, P.J.; Siddell, S.G.; Lefkowitz, E.J.; Mushegian, A.R.; Adriaenssens, E.M.; Dempsey, D.M.; Dutilh, B.E.; Harrach, B.; Harrison, R.L.; Hendrickson, R.C.; et al. Changes to virus taxonomy and the Statutes ratified by the International Committee on Taxonomy of Viruses (2020). *Arch. Virol.* **2020**, *165*, 2737–2748. [CrossRef]
2. Munir, K.; Ashraf, S.; Munir, I.; Khalid, H.; Muneer, M.A.; Mukhtar, N.; Amin, S.; Ashraf, S.; Imran, M.A.; Chaudhry, U.; et al. Zoonotic and reverse zoonotic events of SARS-CoV-2 and their impact on global health. *Emerg. Microbes Infect.* **2020**, *9*, 2222–2235. [CrossRef]
3. V'kovski, P.; Kratzel, A.; Steiner, S.; Stalder, H.; Thiel, V. Coronavirus biology and replication: Implications for SARS-CoV-2. *Nat. Rev. Microbiol.* **2021**, *19*, 155–170. [CrossRef] [PubMed]
4. Andersen, K.G.; Rambaut, A.; Lipkin, W.I.; Holmes, E.C.; Garry, R.F. The proximal origin of SARS-CoV-2. *Nat. Med.* **2020**, *26*, 450–452. [CrossRef] [PubMed]
5. Zhou, P.; Yang, X.L.; Wang, X.G.; Hu, B.; Zhang, L.; Zhang, W.; Si, H.R.; Zhu, Y.; Li, B.; Huang, C.L.; et al. A pneumonia outbreak associated with a new coronavirus of probable bat origin. *Nature* **2020**, *579*, 270–273. [CrossRef] [PubMed]
6. Mathavarajah, S.; Dellaire, G. Lions, tigers and kittens too: ACE2 and susceptibility to COVID-19. *Evol. Med. Public Health* **2020**, *2020*, 109–113. [CrossRef] [PubMed]
7. Wu, L.; Chen, Q.; Liu, K.; Wang, J.; Han, P.; Zhang, Y.; Hu, Y.; Meng, Y.; Pan, X.; Qiao, C.; et al. Broad host range of SARS-CoV-2 and the molecular basis for SARS-CoV-2 binding to cat ACE2. *Cell Discov.* **2020**, *6*, 68. [CrossRef] [PubMed]
8. Alexander, M.R.; Schoeder, C.T.; Brown, J.A.; Smart, C.D.; Moth, C.; Wikswo, J.P.; Capra, J.A.; Meiler, J.; Chen, W.; Madhur, M.S. Predicting susceptibility to SARS-CoV-2 infection based on structural differences in ACE2 across species. *FASEB J.* **2020**, *34*, 15946–15960. [CrossRef]
9. Melin, A.D.; Janiak, M.C.; Marrone, F.; Arora, P.S.; Higham, J.P. Comparative ACE2 variation and primate COVID-19 risk. *Commun. Biol.* **2020**, *3*, 641. [CrossRef] [PubMed]
10. Shi, J.; Wen, Z.; Zhong, G.; Yang, H.; Wang, C.; Huang, B.; Liu, R.; He, X.; Shuai, L.; Sun, Z.; et al. Susceptibility of ferrets, cats, dogs, and other domesticated animals to SARS–coronavirus 2. *Science* **2020**, *368*, 1016–1020. [CrossRef] [PubMed]
11. Halfmann, P.J.; Hatta, M.; Chiba, S.; Maemura, T.; Fan, S.; Takeda, M.; Kinoshita, N.; Hattori, S.-I.; Sakai-Tagawa, Y.; Iwatsuki-Horimoto, K.; et al. Transmission of SARS-CoV-2 in Domestic Cats. *N. Engl. J. Med.* **2020**, *383*, 592–594. [CrossRef]
12. Kim, Y.I.; Kim, S.G.; Kim, S.M.; Kim, E.H.; Park, S.J.; Yu, K.M.; Chang, J.H.; Kim, E.J.; Lee, S.; Casel, M.A.B.; et al. Infection and Rapid Transmission of SARS-CoV-2 in Ferrets. *Cell Host Microbe* **2020**, *27*, 704–709.e2. [CrossRef]
13. Schlottau, K.; Rissmann, M.; Graaf, A.; Schön, J.; Sehl, J.; Wylezich, C.; Höper, D.; Mettenleiter, T.C.; Balkema-Buschmann, A.; Harder, T.; et al. SARS-CoV-2 in fruit bats, ferrets, pigs, and chickens: An experimental transmission study. *Lancet Microbe* **2020**, *1*, e218–e225. [CrossRef]
14. Singh, D.K.; Singh, B.; Ganatra, S.R.; Gazi, M.; Cole, J.; Thippeshappa, R.; Alfson, K.J.; Clemmons, E.; Gonzalez, O.; Escobedo, R.; et al. Responses to acute infection with SARS-CoV-2 in the lungs of rhesus macaques, baboons and marmosets. *Nat. Microbiol.* **2021**, *6*, 73–86. [CrossRef] [PubMed]
15. Munster, V.J.; Feldmann, F.; Williamson, B.N.; van Doremalen, N.; Pérez-Pérez, L.; Schulz, J.; Meade-White, K.; Okumura, A.; Callison, J.; Brumbaugh, B.; et al. Respiratory disease in rhesus macaques inoculated with SARS-CoV-2. *Nature* **2020**, *585*, 268–272. [CrossRef]
16. Chan, J.F.-W.; Zhang, A.J.; Yuan, S.; Poon, V.K.-M.; Chan, C.C.-S.; Lee, A.C.-Y.; Chan, W.-M.; Fan, Z.; Tsoi, H.-W.; Wen, L.; et al. Simulation of the Clinical and Pathological Manifestations of Coronavirus Disease 2019 (COVID-19) in a Golden Syrian Hamster Model: Implications for Disease Pathogenesis and Transmissibility. *Clin. Infect. Dis.* **2020**, *71*, 2428–2446. [CrossRef]
17. Imai, M.; Iwatsuki-Horimoto, K.; Hatta, M.; Loeber, S.; Halfmann, P.J.; Nakajima, N.; Watanabe, T.; Ujie, M.; Takahashi, K.; Ito, M.; et al. Syrian hamsters as a small animal model for SARS-CoV-2 infection and countermeasure development. *Proc. Natl. Acad. Sci. USA* **2020**, *117*, 16587–16595. [CrossRef] [PubMed]
18. Fagre, A.; Lewis, J.; Eckley, M.; Zhan, S.; Rocha, S.M.; Sexton, N.R.; Burke, B.; Geiss, B.; Peersen, O.; Bass, T.; et al. SARS-CoV-2 infection, neuropathogenesis and transmission among deer mice: Implications for spillback to New World rodents. *PLoS Pathog.* **2021**, *17*, e1009585. [CrossRef] [PubMed]
19. Patterson, E.I.; Elia, G.; Grassi, A.; Giordano, A.; Desario, C.; Medardo, M.; Smith, S.L.; Anderson, E.R.; Prince, T.; Patterson, G.T.; et al. Evidence of exposure to SARS-CoV-2 in cats and dogs from households in Italy. *Nat. Commun.* **2020**, *11*, 6231. [CrossRef] [PubMed]
20. Zhang, Q.; Zhang, H.; Gao, J.; Huang, K.; Yang, Y.; Hui, X.; He, X.; Li, C.; Gong, W.; Zhang, Y.; et al. A serological survey of SARS-CoV-2 in cat in Wuhan. *Emerg. Microbes Infect.* **2020**, *9*, 2013–2019. [CrossRef]

21. Oreshkova, N.; Molenaar, R.J.; Vreman, S.; Harders, F.; Oude Munnink, B.B.; Van Der Honing, R.W.H.; Gerhards, N.; Tolsma, P.; Bouwstra, R.; Sikkema, R.S.; et al. SARS-CoV-2 infection in farmed minks, the Netherlands, April and May 2020. *Eurosurveillance* **2020**, *25*, 2001005. [CrossRef]
22. World Organisation for Animal Health. *SARS-CoV-2 in Animals–Situation Report 3*; World Organisation for Animal Health: Paris, France, 2021. Available online: https://www.oie.int/app/uploads/2021/08/sars-cov-2-situation-report-3.pdf (accessed on 5 August 2021).
23. Račnik, J.; Kočevar, A.; Slavec, B.; Korva, M.; Rus, K.R.; Zakotnik, S.; Zorec, T.M.; Poljak, M.; Matko, M.; Rojs, O.Z.; et al. Transmission of SARS-CoV-2 from human to domestic ferret. *Emerg. Infect. Dis.* **2021**, *27*, 2450–2453. [CrossRef] [PubMed]
24. McAloose, D.; Laverack, M.; Wang, L.; Killian, M.L.; Caserta, L.C.; Yuan, F.; Mitchell, P.K.; Queen, K.; Mauldin, M.R.; Cronk, B.D.; et al. From people to panthera: Natural SARS-CoV-2 infection in tigers and lions at the bronx zoo. *MBio* **2020**, *11*, e02220-20. [CrossRef] [PubMed]
25. Karikalan, M.; Chander, V.; Mahajan, S.; Deol, P.; Agrawal, R.K.; Nandi, S.; Rai, S.K.; Mathur, A.; Pawde, A.; Singh, K.P.; et al. Natural infection of Delta mutant of SARS-CoV-2 in Asiatic Lions of India. *Transbound. Emerg. Dis.* **2021**, 1–9. [CrossRef]
26. Bonilla-aldana, D.K.; García-barco, A.; Jimenez-diaz, S.D.; Bonilla-aldana, J.L.; Cardona-trujillo, M.C.; Muñoz-lara, F.; Zambrano, I.; Salas-matta, L.A.; Rodriguez-morales, A.J. SARS-CoV-2 natural infection in animals: A systematic review of studies and case reports and series. *Vet. Q.* **2021**, *41*, 250–267. [CrossRef] [PubMed]
27. Chandler, J.C.; Bevins, S.N.; Ellis, J.W.; Linder, T.J.; Tell, R.M.; Jenkins-Moore, M.; Root, J.J.; Lenoch, J.B.; Robbe-Austerman, S.; DeLiberto, T.J.; et al. SARS-CoV-2 exposure in wild white-tailed deer (Odocoileus virginianus). *Proc. Natl. Acad. Sci. USA* **2021**, *118*, e2114828118. [CrossRef]
28. Tegally, H.; Wilkinson, E.; Lessells, R.J.; Giandhari, J.; Pillay, S.; Msomi, N.; Mlisana, K.; Bhiman, J.N.; von Gottberg, A.; Walaza, S.; et al. Sixteen novel lineages of SARS-CoV-2 in South Africa. *Nat. Med.* **2021**, *27*, 440–446. [CrossRef] [PubMed]
29. Tegally, H.; Wilkinson, E.; Althaus, C.L.; Giovanetti, M.; San, J.E.; Giandhari, J.; Pillay, S.; Naidoo, Y.; Ramphal, U.; Msomi, N.; et al. Rapid replacement of the Beta variant by the Delta variant in South Africa. *medRxiv* **2021**. [CrossRef]
30. Afgan, E.; Baker, D.; Batut, B.; Van Den Beek, M.; Bouvier, D.; Ech, M.; Chilton, J.; Clements, D.; Coraor, N.; Grüning, B.A.; et al. The Galaxy platform for accessible, reproducible and collaborative biomedical analyses: 2018 update. *Nucleic Acids Res.* **2018**, *46*, W537–W544. [CrossRef] [PubMed]
31. Rambaut, A.; Holmes, E.C.; O'Toole, Á.; Hill, V.; McCrone, J.T.; Ruis, C.; du Plessis, L.; Pybus, O.G. A dynamic nomenclature proposal for SARS-CoV-2 lineages to assist genomic epidemiology. *Nat. Microbiol.* **2020**, *5*, 1403–1407. [CrossRef]
32. Hadfield, J.; Megill, C.; Bell, S.M.; Huddleston, J.; Potter, B.; Callender, C.; Sagulenko, P.; Bedford, T.; Neher, R.A. NextStrain: Real-time tracking of pathogen evolution. *Bioinformatics* **2018**, *34*, 4121–4123. [CrossRef] [PubMed]
33. Moshiri, N. ViralMSA: Massively scalable reference-guided multiple sequence alignment of viral genomes. *Bioinformatics* **2021**, *37*, 714–716. [CrossRef]
34. Minh, B.Q.; Schmidt, H.A.; Chernomor, O.; Schrempf, D.; Woodhams, M.D.; Von Haeseler, A.; Lanfear, R.; Teeling, E. IQ-TREE 2: New Models and Efficient Methods for Phylogenetic Inference in the Genomic Era. *Mol. Biol. Evol.* **2020**, *37*, 1530–1534. [CrossRef] [PubMed]
35. Rambaut, A.; Lam, T.T.; Carvalho, L.M.; Pybus, O.G. Exploring the temporal structure of heterochronous sequences using TempEst (formerly Path-O-Gen). *Virus Evol.* **2016**, *2*, vew007. [CrossRef] [PubMed]
36. Drummond, A.J.; Suchard, M.A.; Xie, D.; Rambaut, A. Bayesian phylogenetics with BEAUti and the BEAST 1.7. *Mol. Biol. Evol.* **2012**, *29*, 1969–1973. [CrossRef]
37. Argimón, S.; Abudahab, K.; Goater, R.J.E.; Fedosejev, A.; Bhai, J.; Glasner, C.; Feil, E.J.; Holden, M.T.G.; Yeats, C.A.; Grundmann, H.; et al. Microreact: Visualizing and sharing data for genomic epidemiology and phylogeography. *Microb. Genom.* **2016**, *2*, e000093. [CrossRef] [PubMed]
38. Bartlett, S.L.; Koeppel, K.N.; Cushing, A.C.; Bellon, H.F.; Almagro, V.; Gyimesi, Z.S.; Thies, T.; Hård, T.; Denitton, D.; Fox, K.Z.; et al. Global Retrospective Review of SARS-CoV-2 Infections in Non-Domestic Felids. In Proceedings of the Conference Proceedings: 2021 Joint AAZV/EAZWV Conference, Online, 4 October–5 November 2021; p. 163.

 viruses

MDPI

Review

Nanotechnology as a Shield against COVID-19: Current Advancement and Limitations

Mahendra Rai [1,*], Shital Bonde [1], Alka Yadav [1], Arpita Bhowmik [2], Sanjay Rathod [3], Pramod Ingle [1] and Aniket Gade [1]

[1] Nanobiotechnology Lab., Department of Biotechnology, Sant Gadge Baba Amravati University, Amravati 444 602, Maharashtra, India; shitalbonde@gmail.com (S.B.); nanoalka@gmail.com (A.Y.); pingle23@gmail.com (P.I.); aniketgade@sgbau.ac.in (A.G.)

[2] Faculty of Medicine, Dentistry and Health, The University of Sheffield, Sheffield S10 2TN, UK; arpitabhowmik035@gmail.com

[3] Department of Immunology, University of Pittsburgh, Pittsburgh, PA 15261, USA; SBR21@pitt.edu

* Correspondence: mahendrarai7@gmail.com

Abstract: The coronavirus disease 2019 (COVID-19) caused by severe acute respiratory syndrome coronavirus 2 (SARS-CoV-2) is a global health problem that the WHO declared a pandemic. COVID-19 has resulted in a worldwide lockdown and threatened to topple the global economy. The mortality of COVID-19 is comparatively low compared with previous SARS outbreaks, but the rate of spread of the disease and its morbidity is alarming. This virus can be transmitted human-to-human through droplets and close contact, and people of all ages are susceptible to this virus. With the advancements in nanotechnology, their remarkable properties, including their ability to amplify signal, can be used for the development of nanobiosensors and nanoimaging techniques that can be used for early-stage detection along with other diagnostic tools. Nano-based protection equipment and disinfecting agents can provide much-needed protection against SARS-CoV-2. Moreover, nanoparticles can serve as a carrier for antigens or as an adjuvant, thereby making way for the development of a new generation of vaccines. The present review elaborates the role of nanotechnology-based tactics used for the detection, diagnosis, protection, and treatment of COVID-19 caused by the SARS-CoV-2 virus.

Keywords: COVID-19; SARS-CoV-2; nanotechnology; detection; treatment

Citation: Rai, M.; Bonde, S.; Yadav, A.; Bhowmik, A.; Rathod, S.; Ingle, P.; Gade, A. Nanotechnology as a Shield against COVID-19: Current Advancement and Limitations. *Viruses* **2021**, *13*, 1224. https://doi.org/10.3390/v13071224

Academic Editor: Burtram C. Fielding

Received: 1 June 2021
Accepted: 21 June 2021
Published: 24 June 2021

1. Introduction

Coronavirus disease (COVID-19) or SARS-CoV-2 infection is caused by a virus that belongs to the subfamily Coronavirinae (family: Coronaviridae). The disease emerged at the end of 2019 in the city of Wuhan, China. The virus is spherical, enveloped with spike-like proteins protruding from the virion surface, and has a single-stranded RNA genome. The virus has approximately 79% genomic similarity with the severe acute respiratory syndrome coronavirus (SARS-CoV) and 50% genomic similarity with the Middle East respiratory syndrome coronavirus (MERS-CoV) [1]. SARS-CoV-2 has spread fast worldwide, causing a global pandemic outnumbering the people infected by either SARS-CoV or MERS-CoV since their emergence in 2002 and 2012, respectively. The clinical manifestation of the virus includes fever, dry cough, loss of taste and smell, body pain, anorexia, dyspnea, fatigue, and life-threatening acute respiratory distress syndrome (ARDS) [1]. Although lungs are the primary target of the virus, other systems such as cardiovascular, kidney, liver, central nervous system, and the immune system are also compromised in COVID-19 [2]. As the virus continues to spread in an implacable way causing widespread social, health, and economic disruptions, preventive measures such as social distancing, washing hands, and wearing masks have become pertinent to contain viral transmission. With no official drugs approved for the disease, the current treatment mainly involves symptomatic relief coupled with respiratory support for more severe patients. The heterogeneous nature of the disease and constant mutation in the virus warrants a need for diagnostic tools.

In this regard, nanotechnology is being seriously investigated for its potential in the development of therapeutics, vaccines, diagnostic techniques, and strategies to reduce the healthcare burden. The unique properties of nanoparticles such as their small size, enhanced solubility, better target reachability, improved half-life, reduced side-effects, and surface adaptability are being utilized to bring out a much-needed clinical transformation that could be effective directly against the virus [3,4]. Researchers are now looking into nanotechnology for developing improved assays and nanosensor-based diagnostic techniques, improved delivery of medications, and increased circulation time of the drugs. Thus, nanotechnology seems to hold the potential to bring in innovative alternatives effective against the virus.

2. Risk of Comorbidities in SARS-CoV-2 Infection

Association between the clinical phenotype of COVID-19 and pre-existing chronic co-morbidities is currently poorly described in the literature and is broadly based on small retrospective studies. As per CDC statistics, 80% of COVID-19-related deaths and 47% of hospitalizations occurred in people above 65 years, and co-morbidities played a major contributing factor. The increasing number of patients, poor prognosis in a certain population, and a limited number of medical supplies have overwhelmed the existing healthcare system. Greater understanding of the correlation of risk factors and the disease progression of COVID-19 could help in disease management by personalizing the treatment for improved outcomes. Furthermore, classifying patients into severe and non-severe groups could reduce the healthcare burden.

Co-morbidities including cardiovascular, cerebrovascular, neurological diseases, cancer, and diabetes are closely associated with COVID-19-related ICU admissions and deaths [5,6]. As per a meta-analysis conducted by Khamseh et al., pre-existing cardiovascular diseases increased the risk of severe COVID-19 significantly by 4.8-fold [5]. COVID-19, along with viral pneumonia, has been reported to cause cardiovascular manifestations such as myocardial injury, arrhythmias, myocarditis, and thromboembolism coexisting with the increased amount of cardiac troponin I and c-reactive protein in patients with pre-existing cardiovascular disease [7]. On infection, the virus causes hyperinflammation by over-producing pro-inflammatory cytokines. This hyperinflammation/cytokine storm was seen to further lead to abnormalities in the coagulation system, affecting multiple organs [7]. The potential drug–disease interaction could also contribute to cardiovascular complications from the disease. For example, currently prescribed drugs for COVID-19 such as hydroxychloroquine and azithromycin have pro-arrhythmic effects.

Diabetes, more prevalent in the older population, also increases the risk of developing severe COVID-19. The possible factors may include high glucose levels, impaired metabolic control, low-grade chronic inflammatory state, and impaired coagulation [8]. According to an estimate, 30-day mortality in cancer patients with COVID-19 was 13–33% compared with 0.5–2% in the general population [9]. These statistics were subject to stage, and type of cancer, such as hematological malignancies, were observed to have a worse prognosis over solid tumors. Additionally, in cancer, treatment-related outcomes such as immunosuppression and anemia weakens the body's ability to fight off the disease.

The severity of COVID-19 was closely associated with cerebrovascular and neurological diseases [5]. Along with the respiratory tract, the virus is seen to invade the nervous system, indicated by the loss of smell, taste, and impaired consciousness in the early stage of the infection. An increasing number of studies report delirium, seizures, and encephalopathy as an outcome of the disease, further hinting at its effects in the brain. An independent study conducted by the scientists in the City of Hope, USA, discovered that the ApoE4 gene, known to increase the risk of developing Alzheimer's, was also associated with increased susceptibility and severity of COVID-19 [10]. The ApoE4 gene plays a vital role in modulating the pro-inflammatory activity of the macrophages. Thus, it can be predicted that Alzheimer's patients are at an increased risk of hyperinflammation leading to cytokine storm and severe COVID-19. The damaged blood–brain barrier (BBB) in patients with Alzheimer's and dementia, particularly vascular dementia, predisposes them to the infection, as the brain is more easily accessible to the virus in such patients [11]. Memory impairment makes it even more difficult for these patients to comply with preventive measures, such as wearing masks, sanitizing hands, maintaining social distance, amongst others, increasing their risk of contracting the disease. Early detection is the key in COVID-19, which is not possible in such patients, adding to the risk.

3. Mechanism of Immune Response after Infection of SARS-CoV-2

Generally, the immune system is the best defense mechanism against viruses such as SARS-CoV-2 and other pathogens (bacteria, fungi, and protozoans) by clearing the infections or destroying the virus-infected cells. SARS-CoV-2 primarily comes into circulation via respiratory droplets and additionally through aerosol, direct contact with contaminated surfaces, and fecal–oral transmission [12–14]. The SARS-CoV-2 arrives at the host cells via the respiratory tract, airway, alveolar epithelial cells, vascular endothelial cells, and alveolar macrophages [15–17]. These cells initiate an early virus infection and consequent replication due to their expression of the ACE2 receptor needed for SARS-CoV-2 entry [18]. Contrary to the typical common cold to moderate upper-respiratory illness observed in coronaviruses, the novel SARS-CoV-2 causes severe "flu-like" signs that can proceed to pneumonia, acute respiratory distress (ARDS), renal failure, and in some cases death [19–22].

Once SARS-CoV-2 enters the target cell, the host immune system identifies the whole virus or antigenic parts such as spike proteins and provokes both arms of the immune system (innate and adaptive). Like many other RNA viruses, the recognition of SARS-CoV-2 begins with the detection of its genome by host pattern recognition receptors (PRRs), which signal downstream via recruited adaptor proteins, ubiquitin ligases, and kinases, culminating in transcription factors and the ultimate expression of immune genes, including IFNs, cytokines, and chemokines. The IFN-signaling pathway is frequently a primary target of evasion due to its rapidity and effectiveness in eliminating viral infection. SARS-CoV-2 is highly sensitive to IFN responses and acts at several levels in these pathways to antagonize mammalian immune recognition, intruding with downstream signaling or inhibiting specific IFN-stimulated gene (ISG) products [23]. Severe SARS-CoV-2-infected patients trigger hyperimmune responses with high intensities of inflammatory cytokines/chemokines but not enough antiviral cytokine interferon beta (IFN-β) or interferon lambda (IFN-λ), leading to persistent viremia [24]. SARS-CoV-2 most potently inhibits type I and type III IFN expression in the bronchial epithelial cells of both humans and ferrets [25].

As we know, the adaptive immune response against any viral infection is the key to disease severity; T cells especially are central players in the immune response to viral infection. An enhanced understanding of human T cell-mediated immunity in COVID-19 is vital for optimizing therapeutic and vaccine strategies. The immune system, i.e., the innate and acquired immune response, is activated by SARS-CoV-2 infection. Several studies evaluating the clinical features of SARS-CoV-2-infected patients have reported an incubation time of 4–7 days before the onset of symptoms and an additional 7–10 days before the development of severe COVID-19 [26]. After SARS-CoV-2 entry into the host, a virus attaches to cells expressing ACE2, which facilitates its replication. The viral peptides present through major histocompatibility complex (MHC) class I proteins expressed by antigen-presenting cells (APC) such as dendritic cells (DC) to the cytotoxic CD8+ T cells [27].

Further, these cytotoxic CD8+ T cells activate and expand to initiate virus-specific effector and memory phenotypes. For a quick response, the viral antigens are recognized by APCs, such as DC and macrophages, which present viral epitope to helper CD4+ T cells through MHC-Class-II molecules. By stimulating antibody-producing B cells to produce anti-SARS-CoV-2 antibodies such as anti-SARS-CoV-2 IgM, IgA, and IgG, B cells can directly identify the viruses, get activated by them, and interact with helper CD4+ T cells. The first antibody secretion, i.e., IgM isotype primary virus-specific antibody response, is observed within the first week following symptoms. IgG isotype antibodies response comes after the initial IgM response, which mostly retains a lifelong immunity (Figure 1).

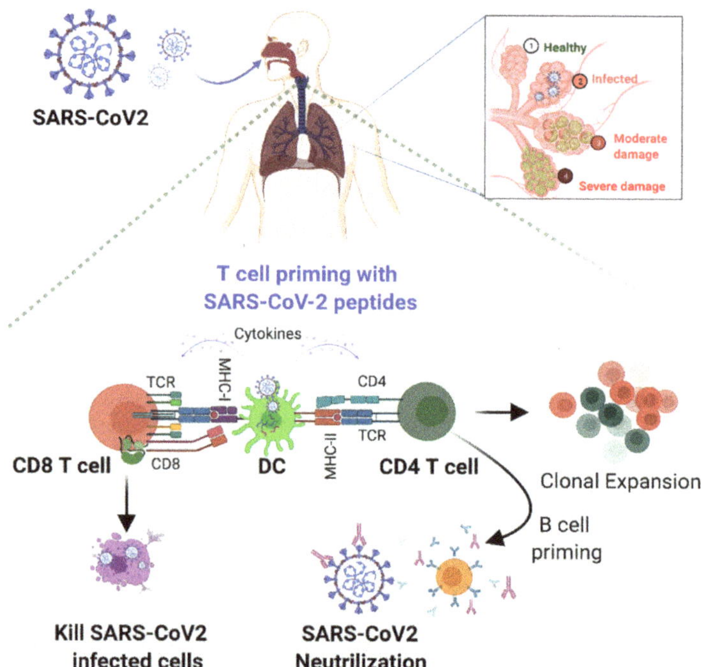

Figure 1. COVID-19 and host immune responses. Following inhalation of SARS-CoV2 into the respiratory tract, the virus traverses deep into the lower lung, where it infects a range of cells expressing its receptor ACE2, including alveolar airway epithelial cells, vascular endothelial cells, and alveolar macrophages. In the innate arm, immune cells primarily recognize the viral RNA by their receptors, such as Toll-like receptors (TLRs) that signal downstream to produce type-I/III interferons (IFNs) and pro-inflammatory mediators as the first line of defense. Furthermore, IFN triggers JAK/STAT signaling to activate interferon stimulating genes (ISGs) to fight SARS-CoV2. In the adaptive arm, the viral peptides are presented through major histocompatibility complex (MHC) class I proteins expressed by dendritic cells (DC) to CD8 T cells; these cells directly kill the virus-infected cells. Further, helper CD4+ T cells are activated through MHC-class II and differentiate B cells into plasma cells (antibody-producing cells) and memory cells. These SARS-CoV2 specific antibodies can neutralize the virus. Overall, both cells play an important role in eradicating SARS-CoV2 from the host.

Although most COVID-19 patients recover from the mild and moderate disease within a week, some individuals develop severe pneumonia in the second week, shadowed by cytokine storm within the third week of the illness. The cytokine storm is a multifaceted network of extreme molecular events integrated by clinical characteristics such as systemic inflammation and multiorgan failure. Cytokine storm is encouraged by the activation of huge numbers of white blood cells, including B cells, T cells, macrophages, dendritic cells, neutrophils, monocytes, NK cells, and resident tissue cells (epithelial and endothelial cells), which secrete high quantities of pro-inflammatory cytokines [28]. Overall, both innate and adaptive systems play an important role in eradicating the SARS-CoV2 from the host.

4. State-of-the-Art of Nanomaterials as Anti-SARS-CoV-2

The recent surge of coronavirus known as SARS-CoV-2 has widely spread across the world; the efficiency of traditional treatment systems also faded due to the emergence of new strains and viral mutations. To overcome the limitations of conventional systems, an improved multidisciplinary approach is needed. Nanomaterials in the form of detection and diagnostic tools, protection equipment, and disinfecting agents can provide much needed protection against SARS-CoV-2.

4.1. *Nanobiosensors*

Although serology-based tests and reverse transcription-polymerase chain reaction (RT-PCR) are routinely used for the detection of COVID-19, there is a need for accuracy and rapidity in diagnosis that can be fulfilled by the use of ultrasensitive nanobiosensors that play a major role in the detection of novel coronavirus. Nanobiosensors provide a rapid, cost-effective, accurate, and miniaturized platform for the detection of SARS-CoV-2 [29].

4.1.1. Affinity-Based Nanobiosensor

Affinity-based nanobiosensors demonstrate the high specificity of bioreceptors, such as antibodies, ssDNA, and aptamers with nanoparticles, which lead to enhanced sensitivity and lower detection limits. Gold nanoparticles, gold nanoislands, graphene, and nanowires are employed for the detection of coronavirus. Gold nanoparticles conjugated with carbon nanotubes improve binding capacity and efficient immobilization matrix. Gold nanoislands are aggregates of gold with a dimension of 20–80 nm and are synthesized by deposition at annealing of gold nanoparticles at elevated temperature for several hours, and these gold nanoislands are also utilized for sensing application [30].

4.1.2. Optical Nanobiosensor

Carbon nanotubes, gold nanoislands, and graphene are majorly used in optical and electrochemical biosensors. Gold nanoislands made of tiny gold nanostructures can be developed with artificially synthesized DNA receptors and complementary RNA sequences of SARS-CoV-2 on a glass substrate. As COVID-19 is a single-stranded RNA virus, the receptor of the nanobiosensor acts as a complementary sequence to the RNA sequence of the coronavirus and detects the virus. LSPR (localized surface plasmon resonance) was used to detect RNA sequence binding to the sensor. After binding of the molecules on the surface of the nanobiosensor, the local infrared index changes and an optical nanobiosensor measures the changes and identifies the presence of RNA strands [28].

Nanobiosensors are used to detect COVID-19. It includes the use of antibodies or cDNA to carefully encapsulate viral RNA. A grapheme-based FET (field effect transistor) device is used for the determination of SARS-CoV-2 viral load in nasopharyngeal swabs of COVID-19 patients. The graphene-based FET nanobiosensor consists of a graphene sheet as the sensing area, transferred to a SiO_2/Si substrate and SARS-CoV-2 spike antibody immobilized on the graphene sheet. The biosensors help detection of SARS-CoV-2 antigen spike even at the concentration of 1 fg/mL in phosphate buffer [30].

4.1.3. Electrochemical Nanobiosensors

Electrochemical sensors are highly sensitive and could be easily miniaturized. Modified electrochemical biosensors in combination with gold nanoparticles show improved applications and can be used for the detection of MERS-CoV. The nanobiosensor is designed with a group of carbon electrode-coated gold nanoparticles. In one study, it was observed that the recombinant spike (S1) protein gets immobilized to gold nanoparticles and competes with the virus particles for binding to the antibody. When there is an absence of virus infection, it binds to the immobilized spike protein. As this nanobiosensor system possesses a group of electrodes, it can be utilized to detect different coronaviruses [31].

Electrochemical nanobiosensors can also be used for the identification of viral nucleic acids [32]. An electrochemical genosensor developed for the detection of SARS was designed by using a monolayer of thiolated oligonucleotides self-assembled on gold nanoparticles-coated carbon electrodes. The oligonucleotide sequences are specific to nucleocapsid protein of SARS, and the viral infection is detected through enzymatic amplification of viral DNA. The nanobiosensor helps the highly sensitive detection of SARS [32]. An electrochemical nanobiosensor fabricated using gold nanoparticles modified with a carbon electrode and recombinant spike protein S1 as biomarker was developed for the detection of MERS-CoVs; however, this technique also holds promise for the detection of coronaviruses. The biosensor was developed using fluorine-doped substrate and gold nanoparticles as a signal amplifier due to its electrical conductivity [33].

4.1.4. Chiral Nanobiosensors

Chiral nanobiosensors provide rapid detection and hence are very useful in distinguishing SARS-CoV-2. Zirconium quantum dots and magnetic nanoparticles in conjugation with coronavirus-specific antibodies bind to the viral target and form magneto plasmonic-fluorescent nanohybrids that could be separated by an external magnet using the optical detection technique. The nanobiosensor showed application in the detection of various virus cultures, including coronaviruses [31].

Ahmed et al. [34] reported a self-assembled technique for the development of a chiral immunosensor using gold nanoparticles and quantum dots. The immunosensor showed detection of virus infection such as adenovirus, avian influenza virus, and coronavirus using blood samples. For the study, virus samples were added to antibody-conjugated chiral gold nanoparticles associated with antibody-conjugated quantum dots. Circular dichroism was used for measuring chiro-optical response.

4.1.5. Nanoimaging System

The Oxford Nanoimaging system can be used for the detection of fluorescently labeled coronaviruses. This system was developed by the scientists from the Department of Physics, University of Oxford. It is an extremely rapid test for the detection of coronavirus. This innovative technology does not require lysis, purification, or amplification process and yields results in 5 min. The technique involves taking direct throat swabs of infected persons and rapid labeling of the virus particles in the sample with short fluorescent DNA strands; the nanoimaging system and machine learning software rapidly detects the virus [35].

4.2. PPE (Personal Protective Equipment) Kits

One of the major reasons for widespread COVID-19 infection is person-to-person contact and the respiratory droplets of the infected person. The healthcare professionals need to use appropriate PPE kits, masks, and gloves to protect themselves from the infection. In such difficult circumstances, nanomaterials prove to be an efficient aid for biological and chemical protection. Nano-engineered facemasks, gloves, and PPE kits provide comfortable, hydrophobic, and antimicrobial activity without altering the texture of a fabric (Figure 2) [36]. PPE kits work as an effective barrier against airborne droplets. The use of nanomaterials with textile fibers can provide antimicrobial properties in textile. For example, nano-silver (AgNPs)-impregnated fabrics have already demonstrated antimicrobial properties. AgNPs-based face masks, smocks, lab coats, hospital curtains, etc. have proved to be highly antimicrobial. In this context, the controlled release of nanoparticles for a longer time can serve in modulating the antiviral properties of the fabric [31].

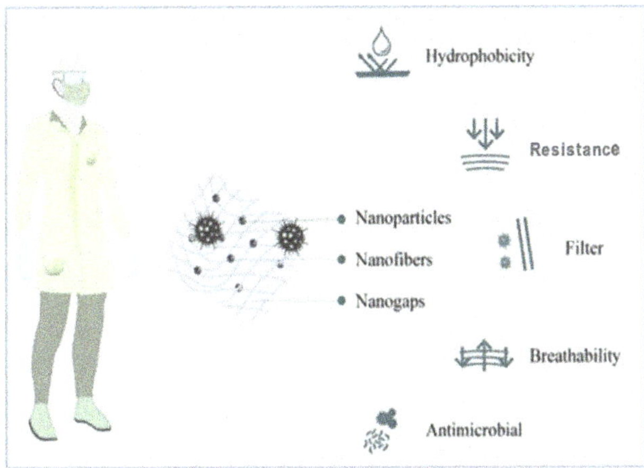

Figure 2. Nanotechnology applications for preparation of PPE Kit [36].

Bhattacharjee et al. [37] reported the use of graphene with silver or copper nanoparticles to enhance the antimicrobial activity of PPE fabric. Graphene on incorporation in a fabric can improve its mechanical strength, antimicrobial property, flame resistance, and the flexibility of the fabric. Metal nanoparticles including silver, copper, and titanium can be associated with graphene to improve its antimicrobial activity, conductivity, and durability. Medical aprons and PPE kits engineered using nanomaterials provide enhanced applications such as hydrophobicity, enhanced antimicrobial activity, and breathability. Hydrophobic nanowhiskers made up of billions of hydrocarbons are extremely small compared to cotton fibers, and they prevent the absorption of droplets. Engineered nanoparticles enhance the surface of textiles and inhibit the growth of pathogenic micro-organisms. Quaternary ammonium salts, polymers, or peptides at the nanoscale level prevent the oxidation of microbial membranes and control their growth [36].

4.3. Nanomasks

One of the most important techniques for prevention against viruses is the use of face masks, as it is crucial for both the infected and non-infected individuals to prevent virus transmission. Various textile products are used for the preparation of facemasks coated with nanoparticles with antiviral properties [38].

Campos et al. [36] also highlighted the use of nanocoated masks for better protection. Nanoparticles do not affect the hydrophobicity and breathability of the fabric. For example, silver and copper metal nanoparticles could be incorporated with different fabrics such as cotton, polyester, polyamide, and cellulose-based fabric to strengthen their use as a filter and also as potential antimicrobial agents. Face masks coated with silver and copper nanoparticle dual-layer coatings have also been designed.

Preliminary studies have demonstrated that silver nanoparticles and silica composite nanocoatings can protect from the lethal effect of SARS-CoV-2. Respiratory face masks incorporating nanoparticles are enhanced owing to the virucidal properties of the nanoparticles. Scientists from the Queensland University of Technology, Australia, have designed a facemask from cellulose nanofibers that can filter particles smaller than 100 nm and is breathable with disposable filter cartridge. Additionally, LIGC Applications Ltd. USA has developed a reusable mask using microporous conductive graphene foam, which traps microbes, and the conduction of electrical charge kills the pathogenic micro-organisms [39]. Nanocellulose nanofibers obtained using plant waste material are claimed to be used for the development of orthogonally aligned nanofiber-based face masks. Nanofibers were produced using insulation and a block electro spinning technique. The orthogonal design of the nanofibers minimized the pressure towards the air filter, enhancing the filtration effect. The nanofiber-based facemasks were water-resistant, had high filtration capacity, and were effective after multiple washes [38].

4.4. Sanitizers and Disinfectants

Viruses are capable of spreading disease and have the capability of becoming a pandemic; however, technological innovations in the field of nanotechnology significantly help in overcoming viruses. Metal nanoparticles such as silver, copper, and titanium show antiviral activity and can be used as an alternative to chemical disinfectants for protection against SARS-CoV-2. [39]. Environmentally friendly, non-irritating nanosilver-based multiuse sanitizer has been introduced using a nanocolloidal technique. The sanitizer shows effective antiviral, antibacterial, and antifungal activities. NanoTechSurface, Italy, has also developed a disinfectant solution based on silver ions and titanium dioxide for disinfecting surfaces contaminated with coronavirus. The nanopolymer-based disinfectant also shows effective antimicrobial activity, is easy to develop, and is cost-efficient, non-inflammable, and biodegradable. This kind of disinfectant has benefits over chemical-based disinfectants as they are biodegradable and do not catch fire. Wero Water Services has designed biopolymer-based disinfectants that are used by the Prague Public Transit Company for sanitization of public transport vehicles [38].

4.5. Antiviral Coatings

Bio-contamination of surfaces and medical devices is a growing concern amid the coronavirus pandemic. The virus-laden respiratory droplets of COVID-19 patients, when exposed in air, deposit on various surfaces and get transmitted to humans; such virus-infected surfaces are known as "fomites" and serve as infectious agents in the transfer of the virus. Traditional disinfecting techniques provide temporary protection, and the bio-burden returns to its original form in a short time span. Non-migratory quaternary ammonium cations (QUATs) and positively charged silver nanoparticles dispensed in polymer matrix can be used for the production of antimicrobial coatings. This coating surface repels oil and water and inactivates coronavirus. It is proposed that silver nanoparticles can inhibit replication of virus nucleotides and inactivate SARS-CoV-2 by interacting with surface spike proteins [40].

Super-hydrophobic nanocoatings could also be used to prevent the transmission of viruses. Copper nanoparticles show antibacterial and antiviral properties and are used to develop super-hydrophobic nanocoatings through the dispersion of nanoparticles in a flexible polymer matrix with the help of a solvent such as acetone. The resultant emulsion can be spray-coated on different surfaces such as doors, knobs, wooden surfaces, and fabrics [40].

Copper and titanium bilayer coatings can be used as nanocoatings over glass surfaces. Silver nanoparticles have also been employed to coat stainless steel surfaces, as most medical devices are made of stainless steel. The synthesis of lysozyme–silver nanoparticles and electrophoretically depositing them on the surface of instruments such as scalpel blades has recently been reported [41]. Erkoc and Uluchan-Karnak [42] demonstrated the use of silver, gold, magnesium oxide, copper oxide, titanium oxide, and zinc oxide nanoparticles to produce coatings with antimicrobial properties. Copper nanoparticles and cardboard materials prevent SARS-CoV-2 infection more efficiently compared with stainless steel and plastic surfaces.

4.6. 3D-Printing

3D printing, also known as additive manufacturing or rapid prototyping, is basically a production technology that utilizes materials such as plastic or metal stacked in 3D layers to create 3D products. 3D printing is mostly used in the field of engineering. It is also used extensively in the healthcare industry. 3D printed face masks, PPE kits, face shields, auxiliary accessories, door openers, and pushbuttons have been designed and offer great opportunities. However, there are several challenges in 3D printing that have to be answered through future research and technology [43,44]. Coronavirus infection can be divided into three stages: asymptomatic incubation period and severe and non-severe symptomatic period. When the defense system of the patient is unable to fight infection, disruption of the tissues occurs, affecting the kidney and intestine and causing inflammation in lungs. 3D printing technology could be used for the production of simple, inexpensive, and structured drug delivery systems using poly (acrylic acid), cellulose acetate, and polyvinyl alcohol (PVA) to prevent infection [43]. Nanomedicines are the future approach in the cure of infectious diseases. The use of metallic nanoparticles, dendrimers, polymer and lipid nanoparticles, quantum dots, and carbon nanotubes have been researched for their applications in nanomedicine. Designing and developing nanomedicines by using 3D print technology will help to satisfy the personal necessity of patients and will also offer biocompatibility (Figure 3) [45]. Thus, 3D printing is a rapid tool for manufacturing PPE to cater to the global demand, which is an alternative to the slow conventional manufacturing processes.

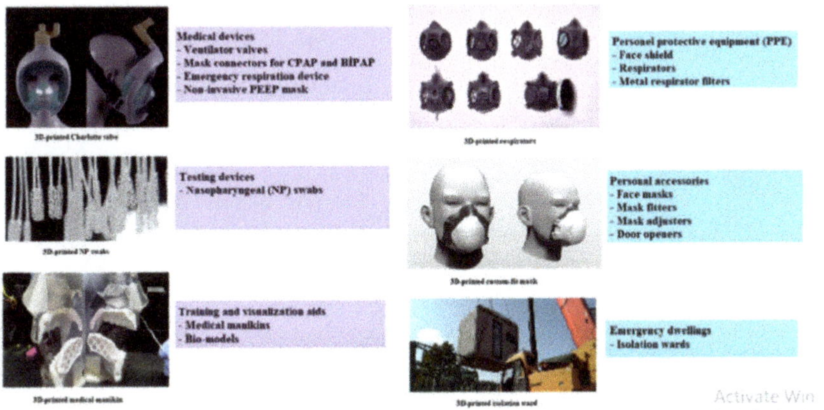

Figure 3. Equipment that could be 3D printed [43].

5. Current Advancements on Nanomedicine: Therapeutics and Vaccine Development

Nanotechnology is opening new therapeutic possibilities of fighting against COVID-19 by enabling new methods of prevention, diagnosis, drug-delivery, and treatment. Nanomedicine is known as the branch of medicine involved in the prevention and cure of various diseases using the nanoscale materials, such as biocompatible nanoparticles [46] and nanorobots [47], for various applications including diagnosis [48], delivery [49], sensing [50]. Nanomedicines have exhibited important features, such as efficient transport through fine capillary blood vessels and lymphatic endothelium, longer circulation duration and blood concentration, higher binding capacity to biomolecules such as endogenous compounds including proteins, higher accumulation in target tissues, reduced inflammatory or immune responses, and oxidative stress in tissues. These features vary from those of conventional medicines dependent on physiochemical properties (e.g., particle surface, size, and chemical composition) of the nanoformulations [49,51,52].

Nanomedicines specifically allow more specific drug targeting and delivery, greater safety, and biocompatibility. The more rapid development of new medicines with wide therapeutic ranges and/or improvement of in vivo pharmacokinetic properties has been reported [52]. The main purpose of nanomedicine is enhanced efficacy and reduced adverse reactions (e.g., toxicity) by altering the efficacy, safety, physicochemical properties, and pharmacokinetic/pharmacodynamic properties of the original drugs [53]. Nanomedicines have greater oral bioavailability. Longer terminal half-life can be predictable in the case of orally administered nanomedicine, which leads to a reduction of administration frequency, dose, and toxicity [53,54]. The nano delivery systems use the nanocarrier for delivering drugs at the target site. Nanocarriers (NCs) shield their load from premature degradation in the biological environment, improve bioavailability, and prolong presence in blood and cellular uptake [55]. Nanoencapsulation is the smart design of nanocarriers and are concerned with the target site and route of administration, attempting to solve the problems faced by therapeutic agents. Effective nanoparticle-based therapy includes FDA-approved lipid systems such as liposomes and micelles [56]. These liposomes and micelles can be loaded with gold or magnetic inorganic nanoparticles [57]. These properties increase the use of inorganic nanoparticles by highlighting drug delivery, imaging, and therapeutics actions. Additionally, nanoparticles help in preventing drugs from being degraded in the gastrointestinal region. They precisely support the sparing delivery of water-soluble drugs to their target location. Formulated nano drugs show higher oral bioavailability, as they display typical uptake mechanisms of absorptive endocytosis [58]. Nanoparticles such as metallic, organic, inorganic, and polymeric nanostructures, as well as dendrimers, micelles, and liposomes, are often considered in designing the target-specific drug delivery systems. Specifically, those drugs having poor solubility with less absorption ability are tagged with these nanoparticles [59]. However, polymeric nanomaterials with diameters ranging from 10 to 1000 nm show the ideal delivery vehicle [60].

Nanotechnological Ways for Vaccine Development

Nanotechnology has caught attention as a potential strategy for the development of a new generation of vaccines, as the nanoparticles serve as a carrier for the antigen and behave as an adjuvant as well in many cases. SARS-COV and MERS treatment and vaccine candidates have not been thoroughly tested and optimized in the past due to considerably lower infection rates than COVID-19, and they have not been noted to have sufficient efficacy. In contrast to SARS or MERS, COVID-19 has been a global threat for more than a year. In research and production, innovative approaches have been recently used [61]. For SARS-CoV, MERS-CoV has been used to introduce nanotechnology into vaccines and therapeutic research on several occasions. Virus-like particles (VLPs) have recently been reported to be suitable for the development of vaccines or treatments for MERS-CoV infection symptoms [62]. Nano-sized VLPs can be delivered through the lymphatic system and capillaries in a better way than other small vaccines because they have the characteristic functions of viruses [63,64]. Additionally, they also reduce the systemic inflammatory response, and have the advantage of being able to enter cells very easily, much like the virus itself. Moreover, delivering a large number of antigens improves the antigen-presenting cell's efficiency. As a result, the T cell receptor recognizes the synthesized complex, increasing the vaccine's immunogenicity and efficacy [64].

VLPs that enter into the host cell are involved in B cell activation and immune system stimulation. Nano-sized VLPs have been shown to effectively overcome viruses by increasing immune response in animal experiments [65,66]. These findings were investigated for the S protein, which is found in both MERS-CoV and SARS-CoV, and hence, they can be used to effectively treat SARS-CoV-2 infection. The advantage of the present SARS-CoV-2 vaccines (approved and in the development process) is that they can be used for drug and gene delivery. The liposomes are suited to deliver nucleic acid [67].

6. Nanotechnology-Based Approaches in Preclinical and Clinical Studies: In Vitro and In Vivo

6.1. Nano-Based Approaches in Pre-Clinical Studies

COVID-19 immune-based preclinical therapeutic approaches such as virus-binding molecules; inhibitors of specific enzymes involved in viral replication and transcription; small-molecule inhibitors of helicase, proteases, or other proteins critical for the virus survival; host cell protease; endocytosis inhibitors; and siRNA inhibitors are all potential therapeutic options for SARS-CoV-2 [68]. The effects induced by monoclonal antibodies (mAb) in COVID-19 patients may also improve the development of vaccines and increasingly specific diagnostics [69]. Moreover, every single one of these tools needs to be assessed regarding clinical efficacy and safety before treating infected patients.

6.2. Nano-Based Approaches in Clinical Studies

Currently, nanotechnology-based formulations have been developed and commercialized for common viral infections. Several companies are moving away from conventional treatment and prevention strategies and switching over to nanotechnology for developing various types of vaccines and therapeutics, e.g., examethasones, a COVID-19 therapeutic agent that has been introduced via various nanoformulations in the treatment of COVID-19. Completing phase 3 clinical trials of Pfizer's liposomal mRNA vaccine (BNT162b) can be considered a significant achievement in nanomedicine [70].

mRNA- and DNA-based vaccines would have little efficacy without nanomedicine components. According to recent research, nanomaterials may effectively inactivate SARS-CoV-2 virus, as nanomaterials have been used to inhibit viruses of other members of the Coronaviridae family [71]. Many vaccine candidates under development for the SARS-CoV-2 vaccine have safety and efficacy in the clinical and pre-clinical stages [72]. ModernaTX, Inc. used lipid nanoparticles (LNP) to encapsulate mRNA-1273, which encodes the full-length SARS-CoV-2 S protein (NCT04283461). Cells that express this viral protein will be able to present SARS-CoV-2 antigen to T cells, eliciting an immune response against the virus [73], which helps in preventing premature degradation during drug delivery. Other clinical studies are testing diverse anti-inflammatory agents to reduce lung inflammation (pneumonia), the leading cause of death in COVID-19 patients. These contain antibodies targeting inflammatory factors such as IL-6 and complement protein C5, or the CD24Fc conjugate that blocks TLR activation. There are two clinical studies that include the anti-angiogenic drug bevacizumab (anti-VEGF mAb) for reduction of lung oedema. A new antibody in clinical development is meplazumab, which blocks the binding of SARS-CoV-2 S protein to CD147 molecule on human cells, thereby reducing the virus's infection ability. Additional immunosuppressive agents are also being tested, such as the JAK1/JAK2 inhibitor baricitinib and the antimalarial drug hydroxychloroquine sulfate. While optimal treatment regimens are still under study, different dosing and schedules are being reported by clinicians [74]. The immune response by using lipid NPs-mediated drug delivery and mRNA vaccine is shown in Figure 4.

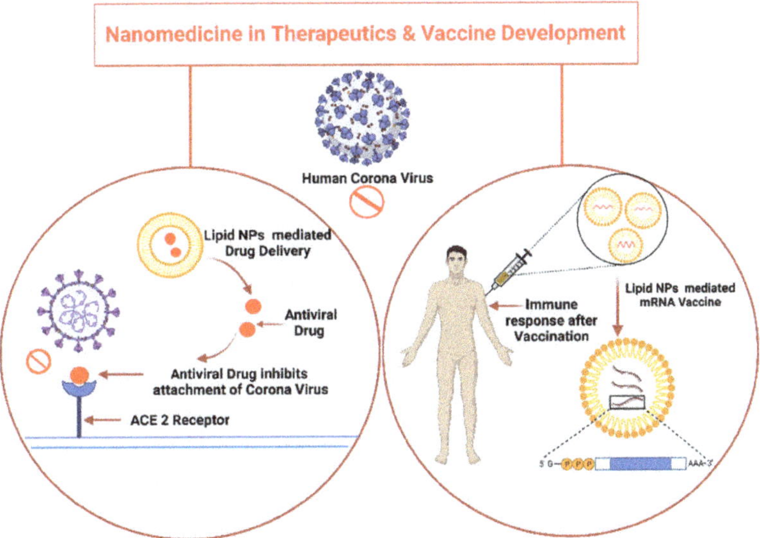

Figure 4. Nanomedicine in Therapeutics and Vaccine Development.

7. Future Perspectives to Tackle COVID-19 Using Nanotechnology

COVID-19 has introduced the scientific community to a global challenge it has perhaps never had to face before. However, it has also taught scientists and the population at large that this kind of situation could occur again. Cutting-edge tools, notably nanotechnology, should be solidly developed to tackle SARS-CoV2 infection. Nanoparticle-based medicine is a very effective tool with the potential to reduce the burden of illness. Nanoparticles that are much smaller than a micrometer have received exceptional attention in managing COVID-19 disease caused by SARS-CoV2 due to their distinctive properties (suitable size, simple preparation, minimal cost, effortless modification, etc.). Nanotechnology-based approaches for combating COVID-19 include the innovation of tools for speedy, precise, and sensitive diagnosis of SARS-CoV2 infection, production of efficient disinfectants, efficient delivery of mRNA-based vaccines into human cells, and delivery of antiviral drugs into the host. Nanotechnology is being geared up for implementation in the fight against SARS-CoV2 infection in a wide range of areas, as shown in Figure 5.

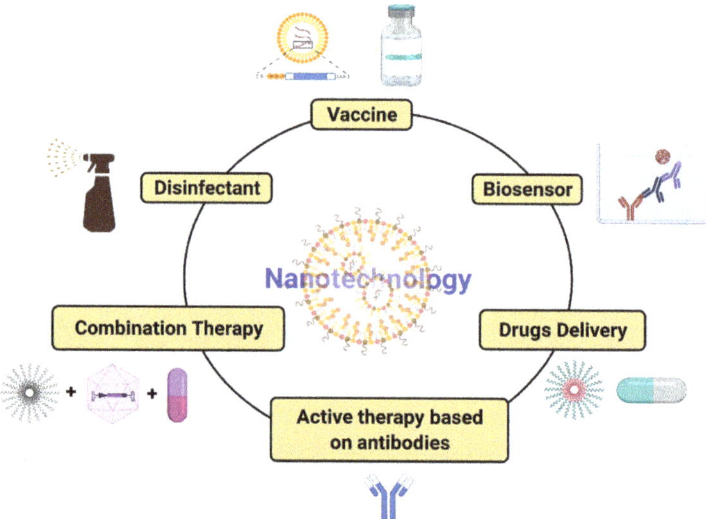

Figure 5. Potential strategies to tackle SARS-CoV-2 utilizing nanotechnology.

Despite the recent progress and intensive studies on nanotechnology-based tools to mitigate COVID-19, there are several important challenges remaining to be addressed when attempting to tackle COVID-19: (i) early, portable, rapid, exceedingly sensitive, and reasonable development of diagnostic kits; (ii) potential use of nanomaterials to avoid the conventional restriction associated with antiviral drugs; (iii) nanoparticle-based vaccine development to fight against SARS-CoV-2 and other pathogens; (iv) combination therapy by utilizing nanoparticles as a delivery system; (v) development of nanobiosensors for rapid and early detection of viruses; and (vi) nanomaterial-based disinfectant agents that can kill pathogens.

Some of the drawbacks associated with nanoparticles, such as cell toxicity, genotoxicity fibrosis, inflammation, immunotoxicity, and oxidative stress, are key issues to be solved before their use with patients. We anticipate that many advances will soon be accomplished in COVID-19 diagnosis, treatment, and therapy using nanotechnology-based strategies. Nanotechnology-based tools will probably be utilized in the treatment of COVID-19 and emerging pathogens. This can be achieved by nanotechnology-based therapeutic antibodies or mRNA- or protein-based vaccines, which specifically deliver the active drugs/epitopes to the host's targeted organs and provide rapid detection of these viruses. Finally, the greatest challenge will be transferring nanomaterial technology to actual clinical applications and the feasibility of production on a large scale.

8. Challenges and Limitations of Nanotechnology in COVID-19

Nanotechnology-based systems, despite their benefits face numerous obstacles before they can be safely introduced to the market. Scalability and production costs are the most common issues, as are intellectual and regulatory properties and potential toxicity and environmental effects of these systems [75]. However, some bottlenecks in nanotechnology applications must be addressed before they are widely adopted in the healthcare system. The major task will be to ensure the safety of nanomaterial via in vitro studies of their biocompatibility. The fate of nanomaterials can be changed into the body when they travel through blood due to the formation of protein corona [76]. Hence, in vivo studies need to be executed carefully to better understand the toxicity of nanoparticles in the body [77]. Because of limitations, generic protocols have been employed for categorization at an early stage of research and development that minimize the chances of failures in terms of clinical translation of nanotechnology-based therapy [78]. To overcome other limitations, a closer collaboration between regulatory agencies, scientific experts in material science, pharmacology, and toxicology is required. The possible toxicity is the main concern of their use in medicine. Thus, not only the positive results of the use of nanoparticles but the appearance of unpredictable results of their action on the human body should also be investigated and scrutinized [79].

The toxicity of nanoparticles is associated with their distribution in the bloodstream and lymph streams as well as with their ability to penetrate almost all cells, tissues, and organs, as well as their ability to interact with different macromolecules. The toxicity of nanoparticles can alter the structure and functioning of organs. Nanoparticle toxicity highly depends on their physical and chemical properties, such as shape, size, surface charge, and the chemical composition of the core and shell. Several types of nanoparticles are not recognized by the body's defense system, which may lead to the accumulation of nanoparticles in organs and tissues, leading to high toxicity or lethality. The solution is to design nanoparticles with a decreased toxicity compared with the traditional nanoparticles that are available. More advanced methods and research should be developed for studying nanoparticles' toxicity and to analyze different pathways and mechanisms of toxicity at the molecular level [80]. Campos et al. investigated the design of nanoparticles that have small or no negative effects and concluded that it is impossible to do so unless all qualitative and quantitative physical and chemical properties of nanoparticles are systematically taken into consideration and a relevant experimental model for estimating their influence on biological systems is available [36].

9. Conclusions

Nanotechnology has emerged as a potential approach to the diagnosis, protection, drug delivery, and development of therapeutic strategies for controlling global pandemics such as COVID-19. Nanoparticles can serve as ideal drug carriers for pulmonary drug delivery, can be used for early and rapid detection of viruses, as part of effective treatments, and are used for nanovaccine preparation by serving as adjuvants that enhance immunogenicity and protect antigens against degradation. The functionalization of nanoparticles with versatile biomolecules and motifs that target SARS-CoV-2 would effectively develop the strategy for treatment and detection. Moreover, there are additional advantages of using nanoparticles with COVID-19 patients, particularly in hospital-acquired co-infections and superinfections caused by bacteria (*Streptococcus pneumoniae, Staphylococcus aureus, Escherichia coli, Pseudomonas aeruginosa*) and fungi (*Aspergillus* spp., *Candida* spp, *Mucor* spp., etc.). However, before using nanoparticles, their toxicity should be evaluated on experimental animals. In addition, dose dependency, the route of administration, biodistribution, and biodegradability of nanoparticles should also be considered. Finally, considering the grave situation caused by the COVID-19 pandemic, it is believed that the existing conventional platform needs to be replaced with new and emerging nanobiotechnological strategies for research on the pandemic that is caused by COVID-19 as well as in research on other related viruses.

Author Contributions: Designed and editing by M.R. and all other authors contributed in writing, editing, and final approval of this review. All authors have read and agreed to the published version of the manuscript.

Funding: This research received no external funding.

Institutional Review Board Statement: Not applicable.

Informed Consent Statement: Not applicable.

Data Availability Statement: Exclude this statement.

Conflicts of Interest: Authors declare no conflict of interest.

References

1. Lu, R.; Zhao, X.; Li, J.; Niu, P.; Yang, B.; Wu, H.; Wang, W.; Song, H.; Huang, B.; Zhu, N.; et al. Genomic characterisation and epidemiology of 2019 novel coronavirus: Implications for virus origins and receptor binding. *Lancet* **2020**, *395*, 565–574. [CrossRef]
2. Cevik, M.; Kuppalli, K.; Kindrachuk, J.; Peiris, M. Virology, transmission, and pathogenesis of SARS-CoV-2. *BMJ* **2020**, *371*, m3862. [CrossRef] [PubMed]
3. Tang, Z.; Zhang, X.; Shu, Y.; Guo, M.; Zhang, H.; Tao, W. Insights from nanotechnology in COVID-19 treatment. *Nano Today* **2020**, *36*, 101019. [CrossRef] [PubMed]
4. Rangayasami, A.; Kannan, K.; Murugesan, S.; Radhika, D.; Sadasivuni, K.K.; Reddy, K.R.; Raghu, A.V. Influence of nanotechnology to combat against COVID-19 for global health emergency: A review. *Sens. Int.* **2021**, *2*, 100079. [CrossRef]
5. Honardoost, M.; Janani, L.; Aghilia, R.; Emamia, Z.; Khamseh, M.E. The Association between Presence of Comorbidities and COVID-19 Severity: A Systematic Review and Meta-Analysis. *Cerebrovasc. Dis.* **2021**, *50*, 132–140. [CrossRef] [PubMed]
6. Biswas, M.; Rahaman, S.; Biswas, T.K.; Haque, Z.; Ibrahim, B. Association of sex, age, and comorbidities with mortality in COVID-19 patients: A systematic review and meta-analysis. *Intervirology* **2021**, *64*, 36–47. [CrossRef] [PubMed]
7. Nishiga, M.; Wang, D.W.; Han, Y.; Lewis, D.B.; Wu, J.C. COVID-19 and cardiovascular disease: From basic mechanisms to clinical perspectives. *Nat. Rev. Cardiol.* **2020**, *17*, 543–558. [CrossRef] [PubMed]
8. Abu-Farha, M.; Al-Mulla, F.; Thanaraj, T.A.; Kavalakatt, S.; Ali, H.; Ghani, M.A.; Abubaker, J. Impact of diabetes in patients diagnosed with COVID-19. *Front. Immunol.* **2020**, *11*, 576818. [CrossRef] [PubMed]
9. Grivas, P.; Khaki, A.R.; Wise-Draper, T.M.; French, B.; Hennessy, C.; Hsu, C.-Y.; Shyr, Y.; Li, X.; Choueiri, T.K.; Painter, C.A.; et al. Association of clinical factors and recent anticancer therapy with COVID-19 severity among patients with cancer: A report from the COVID-19 and Cancer Consortium. *Ann. Oncol* **2021**, *31*, 787–800. [CrossRef]
10. Wang, C.; Zhang, M.; Garcia, G., Jr.; Tian, E.; Cui, Q.; Chen, X.; Sun, G.; Wang, J.; Arumugaswami, V.; Shi, Y. ApoE-isoform-dependent SARS-CoV-2 neurotropism and cellular response. *Cell. Stem. Cell.* **2021**, *28*, 331–342. [CrossRef]
11. Wang, Q.Q.; Davis, P.B.; Gurney, M.E.; Xu, R. COVID-19 and dementia: Analyses of risk, disparity, and outcomes from electronic health records in the US. *Alzheimer's Dement* **2021**, alz.12296. [CrossRef]
12. Yu, I.T.S.; Li, Y.; Wong, T.W.; Tam, W.; Chan, A.T.; Lee, J.H.W.; Leung, D.Y.C.; Tommy Ho, B.S. Evidence of Airborne Transmission of the Severe Acute Respiratory Syndrome Virus. *N. Engl. J. Med.* **2004**, *350*, 1731–1739. [CrossRef]
13. Otter, J.A.; Donskey, C.; Yezli, S.; Douthwaite, S.; Goldenberg, S.D.; Weber, D.J. Transmission of SARS and MERS coronaviruses and influenza virus in healthcare settings: The possible role of dry surface contamination. *J. Hosp. Infect* **2016**, *92*, 235–250. [CrossRef]
14. Li, Q.; Hu, C.; Haerter, R.; Englert, U. Isosteric molecules in non-isomorphous structures: A new route to new structures. The example of 9,10-dihaloanthracene. *CrystEngComm* **2004**, *6*, 83–86. [CrossRef]
15. Kuba, K.; Imai, Y.; Rao, S.; Gao, H.; Guo, F.; Guan, B.; Huan, Y.; Yang, P.; Zhang, Y.; Deng, W.; et al. A crucial role of angiotensin converting enzyme 2 (ACE2) in SARS coronavirus–induced lung injury. *Nat. Med.* **2005**, *11*, 875–879. [CrossRef]
16. Jia, H.P.; Look, D.C.; Shi, L.; Hickey, M.; Pewe, L.; Netland, J.; Farzan, M.; Wohlford-Lenane, C.; Perlman, S.; McCray, P.B., Jr. ACE2 receptor expression and severe acute respiratory syndrome coronavirus infection depend on differentiation of human airway epithelia. *J. Virol.* **2005**, *79*, 14614–14621. [CrossRef]
17. Hamming, I.; Timens, W.; Bulthuis, M.L.C.; Lely, A.T.; Navis, G.J.; Van Goor, H. Tissue distribution of ACE2 protein, the functional receptor for SARS coronavirus. A first step in understanding SARS pathogenesis. *J. Pathol.* **2004**, *203*, 631–637. [CrossRef]
18. Ziegler, C.G.K.; Allon, S.J.; Nyquist, S.K.; Mbano, I.M.; Miao, V.N.; Tzouanas, C.N.; Tzouanas, C.N.; Cao, Y.; Yousif, A.S.; Bals, J.; et al. SARS-CoV-2 receptor ACE2 is an interferon-stimulated gene in human airway epithelial cells and is detected in specific cell subsets across tissues. *Cell* **2020**, *181*, 1016–1035. [CrossRef]

19. Wang, D.; Hu, B.; Hu, C.; Zhu, F.; Liu, X.; Zhang, J.; Wang, B.; Xiang, H.; Cheng, Z.; Xiong, Y.; et al. Clinical Characteristics of 138 Hospitalized Patients With 2019 Novel Coronavirus–Infected Pneumonia in Wuhan, China. *JAMA* **2020**, *323*, 1061–1069. [CrossRef]
20. Chan, J.F.W.; Yuan, S.; Kok, K.H.; To, K.K.W.; Chu, H.; Yang, J.; Xing, F.; Nurs, J.L.B.; Yip, C.C.-Y.; Poon, R.W.-S.; et al. A familial cluster of pneumonia associated with the 2019 novel coronavirus indicating person-to-person transmission: A study of a family cluster. *Lancet* **2020**, *395*, 514–523. [CrossRef]
21. Ksiazek, T.G.; Erdman, D.; Goldsmith, C.S.; Zaki, S.R.; Peret, T.; Emery, S.; Tong, S.; Urbani, C.; Comer, J.A.; Lim, M.P.H.W.; et al. A novel coronavirus associated with severe acute respiratory syndrome. *N. Engl. J. Med.* **2003**, *348*, 1953–1966. [CrossRef]
22. Guery, B.; Poissy, J.; elMansouf, L.; Séjourné, C.; Ettahar, N.; Lemaire, X.; Vuotto, F.; Goffard, A.; PharmD, S.B.; Enouf, V.; et al. Clinical features and viral diagnosis of two cases of infection with Middle East Respiratory Syndrome coronavirus: A report of nosocomial transmission. *Lancet* **2013**, *381*, 2265–2272. [CrossRef]
23. Totura, A.L.; Baric, R.S. SARS coronavirus pathogenesis: Host innate immune responses and viral antagonism of interferon. *Curr. Opin. Virol.* **2012**, *2*, 264–275. [CrossRef]
24. Hadjadj, J.; Yatim, N.; Barnabei, L.; Corneau, A.; Boussier, J.; Smith, N.; Péré, H.; Charbit, B.; Bondet, V.; Chenevier-Gobeaux, C.; et al. Impaired type I interferon activity and exacerbated inflammatory responses in severe Covid-19 patients. *MedRxiv* **2020**, *369*, 718–724.
25. Blanco-Melo, D.; Nilsson-Payant, B.E.; Liu, W.C.; Uhl, S.; Hoagland, D.; Møller, R.; Jordan, T.X.; Oishi, K.; Panis, M.; Sachs, D.; et al. Imbalanced host response to SARS-CoV-2 drives development of COVID-19. *Cell* **2020**, *181*, 1036–1045.e9. [CrossRef]
26. Huang, C.; Wang, Y.; Li, X.; Ren, L.; Zhao, J.; Hu, Y.; Zhang, P.L.; Fan, G.; Xu, J.; Gu, X.; et al. Clinical features of patients infected with 2019 novel coronavirus in Wuhan, China. *Lancet* **2020**, *395*, 497–506. [CrossRef]
27. Jansen, J.M.; Gerlach, T.; Elbahesh, H.; Rimmelzwaan, G.F.; Saletti, G. Influenza virus-specific CD4+ and CD8+ T cell-mediated immunity induced by infection and vaccination. *J. Clin. Virol.* **2019**, *119*, 44–52. [CrossRef]
28. Behrens, E.M.; Koretzky, G.A. Cytokine storm syndrome: Looking toward the precision medicine era. *Arthritis Rheumatol.* **2017**, *69*, 1135–1143. [CrossRef]
29. Alhalaili, B.; Popescu, I.N.; Kamoun, O.; Alzubi, F.; Alawadhia, S.; Vidu, R. Nanobiosensors for the detection of novel coronavirus 2019-nCoV and other pandemic/Epidemic Respiratory viruses: A review. *Sensors* **2020**, *20*, 6591. [CrossRef]
30. Antiochia, R. Nanobiosensors as new diagnostic tools for SARS, MERS and COVID-19: From past to perspectives. *Microchim. Acta* **2020**, *187*, 639. [CrossRef]
31. Jindal, S.; Gopinath, P. Nanotechnology based approaches for combating COVID-19 viral infection. *Nano Express* **2020**, *1*, 022003. [CrossRef]
32. Martinez-Paredes, G.; Gonzalez-Garcia, M.B.; Costa-Garcia, A. Genosensor for SARS virus detection based on gold nanostructured screen-printed carbon electrodes. *Electroanal* **2009**, *21*, 379–385. [CrossRef]
33. Iravani, S. Nano-and biosensors for the detection of SARS-CoV-2: Challenges and opportunities. *Mater. Adv.* **2020**, *1*, 3092. [CrossRef]
34. Ahmed, S.R.; Nagy, E.; Neethirajan, S. Self-assembled star-shaped chiroplasmonic gold nanoparticles for an ultrasensitive chiroimmunosensor for viruses. *RSC Adv.* **2017**, *7*, 40849–40857. [CrossRef]
35. Shiaelis, N.; Tometzki, A.; Peto, L.; McMahon, A.; Hepp, C.; Bickerton, E.; Favard, C.; Muriaux, D.; Andersson, M.; Oakley, S.; et al. Virus detection and identification in minutes using single-particle imaging and deep learning. *MedRxiv* **2021**. [CrossRef]
36. Campos, E.V.R.; Pereira, A.E.S.; de Oliveira, J.L.; Carvalho, L.B.; Guilger-Casagrande, M.; de Lima, R.; Fraceto, L.F. How can nanotechnology help to combat COVID-19? Opportunities and urgent need. *J. Nanobio.* **2020**, *18*, 125. [CrossRef]
37. Bhattacharjee, S.; Joshi, R.K.; Chughtai, A.A.; Macintyre, C.R. Graphene modified multifunctional personal protective clothing. *Adv. Mater. Interfaces* **2019**, *6*, 1900622. [CrossRef]
38. Chaudhary, V.; Royal, A.; Chavali, M.; Yadav, S.K. Advancements in Research and development to combat COVID-19 using Nanotechnology. *Nanotech. Environ. Eng.* **2021**, *6*, 8. [CrossRef]
39. Talebian, S.; Wallace, G.G.; Schroeder, A.; Stellacci, F.; Conde, J. Nanotechnology-based disinfectants and sensors for SARS-CoV-2. *Nat. Nanotechnol.* **2020**, *15*, 615–628. [CrossRef]
40. Shirvanimoghaddam, K.; Akbari, M.K.; Yadav, R.; Al-Tamimi, A.K.; Naebe, M. Fight against COVID-19: The case of antiviral surfaces. *APL Mater.* **2021**, *9*, 031112. [CrossRef]
41. Basak, S.; Packirisamy, G. Nano-based antiviral coatings to combat viral infections. *Nanostruct. Nanoobjects* **2020**, *24*, 100620.
42. Erkoc, P.; Ulucan-Karnak, F. Nanotechnology-based antimicrobial and antiviral surface coating strategies. *Prosthesis* **2021**, *3*, 25–52. [CrossRef]
43. Aydin, A.; Demirtas, Z.; Ok, M.; Erkus, H.; Cebi, G.; Uysal, E.; Gunduz, O.; Ustundag, C.B. 3D printing in the battle against COVID-19. *Emerg. Mater.* **2021**, *4*, 363–386. [CrossRef]
44. Oladapo, B.I.; Ismail, S.O.; Afolalu, T.D.; Olawade, D.B.; Zahedi, M. Review on 3D printing: Fight against COVID-19. *Mat. Chem. Phys.* **2021**, *258*, 123943. [CrossRef]
45. Jain, K.; Shukla, R.; Yadav, A.; Ujjwal, R.R.; Flora, S.J.S. 3D printing in development of Nanomedicine. *Nanomater* **2021**, *11*, 420. [CrossRef]
46. McNamara, K.; Tofail, S.A. Nanosystems: The use of nanoalloys, metallic, bimetallic, and magnetic nanoparticles in biomedical applications. *Phys. Chem. Chem. Phys.* **2015**, *17*, 27981–27995. [CrossRef]

47. Saadeh, Y.; Vyas, D. Nanorobotic applications in medicine: Current proposals and designs. *Am. J. Robot. Surg.* **2014**, *1*, 4–11. [CrossRef]
48. Oliveira, O.N., Jr.; Iost, R.M.; Siqueira, J.R., Jr.; Crespilho, F.N.; Caseli, L. Nanomaterials for diagnosis: Challenges and applications in smart devices based on molecular recognition. *ACS Appl. Mater. Interfaces* **2014**, *6*, 14745–14766. [CrossRef]
49. De Jong, W.H.; Borm, P.J. Drug delivery and nanoparticles: Applications and hazards. *Int. J. Nanomed.* **2008**, *3*, 133. [CrossRef]
50. Holzinger, M.; Le Goff, A.; Cosnier, S. Nanomaterials for biosensing applications: A review. *Front. Chem.* **2014**, *2*, 63. [CrossRef]
51. Liu, W.; Yang, X.L.; Ho, W.S. Preparation of uniform-sized multiple emulsions and micro/nano particulates for drug delivery by membrane emulsification. *J. Pharm. Sci.* **2011**, *100*, 75–93. [CrossRef]
52. Onoue, S.; Yamada, S.; Chan, H.K. Nanodrugs: Pharmacokinetics and safety. *Int. J. Nanomed.* **2014**, *20*, 1025–1037. [CrossRef]
53. Dawidczyk, C.M.; Kim, C.; Park, J.H.; Russell, L.M.; Lee, K.H.; Pomper, M.G.; Searson, P.C. State-of-the-art in design rules for drug delivery platforms: Lessons learned from FDA-approved nanomedicines. *J. Control. Release* **2014**, *10*, 133–144. [CrossRef]
54. Charlene, M.D.; Luisa, M.R.; Peter, C.S. Nanomedicines for cancer therapy: State-of-the-art and limitations to pre-clinical studies that hinder future developments. *Front. Chem.* **2014**, *2*, 69.
55. Kumari, A.; Yadav, S.K. Cellular interactions of therapeutically delivered nanoparticles. *Exp. Opinion Drug. Deliv.* **2011**, *8*, 141–151. [CrossRef]
56. Shi, X.; Sun, K.; Baker, J.R., Jr. Spontaneous formation of functionalized dendrimer-stabilized gold nanoparticles. *J. Phys. Chem. C.* **2008**, *112*, 8251–8258. [CrossRef]
57. Park, S.H.; Oh, S.G.; Mun, J.Y.; Han, S.S. Loading of gold nanoparticles inside the DPPC bilayers of liposome and their effects on membrane fluidities. *Coll. Surf. B* **2006**, *48*, 112–118. [CrossRef]
58. Patra, J.K.; Das, G.; Fraceto, L.F.; Campos, E.V.; del Pilar Rodriguez-Torres, M.; Acosta-Torres, L.S.; Diaz-Torres, L.A.; Grillo, R.; Swamy, M.K.; Sharma, S.; et al. Nano based drug delivery systems: Recent developments and future prospects. *J. Nanobiotechnol.* **2018**, *16*, 71. [CrossRef]
59. Behera, S.; Rana, G.; Satapathy, S.; Mohanty, M.; Pradhan, S.; Panda, M.K.; Ningthoujam, R.; Hazarika, B.N.; Singh, Y.D. Biosensors in diagnosing COVID-19 and recent development. *Sens. Internat.* **2020**, *1*, 100054. [CrossRef]
60. Bonifácio, B.V.; da Silva, P.B.; dos Santos Ramos, M.A.; Negri, K.M.S.; Bauab, T.M.; Chorilli, M. Nanotechnology-based drug delivery systems and herbal medicines: A review. *Int. J. Nanomed.* **2014**, *9*, 1–15.
61. Chakravarty, M.; Vora, A. Drug Delivery and Translational Research. *Drug. Deliv. Transl. Res.* **2014**, 1–10.
62. Reddy, S.T.; Van der Vlies, A.J.; Simeoni, E.; Angeli, V.; Randolph, G.J.; O'Neil, C.P.; Lee, L.K.; Swartz, M.A.; Hubbell, J.A. Exploiting lymphatic transport and complement activation in nanoparticle vaccines. *Nat. Biotechnol.* **2007**, *25*, 1159–1164. [CrossRef] [PubMed]
63. Reddy, S.T.; Berk, D.A.; Jain, R.K.; Swartz, M.A. A sensitive in vivo model for quantifying interstitial convective transport of injected macromolecules and nanoparticles. *J. Appl. Physiol.* **2006**, *101*, 1162–1169. [CrossRef] [PubMed]
64. Bachmann, M.F.; Jennings, G.T. Vaccine delivery: A matter of size, geometry, kinetics and molecular patterns. *Nat. Rev. Immunol.* **2010**, *10*, 787–796. [CrossRef]
65. Ma, C.Q.; Wang, L.L.; Tao, X.R.; Zhang, N.; Yang, Y.; Tseng, C.-T.K.; Li, F.; Zhou, Y.; Jiang, S.; Dua, L. Searching for an ideal vaccine candidate among different MERS coronavirus receptor-binding fragments-The importance of immune focusing in subunit vaccine design. *Vaccine* **2014**, *32*, 6170–6176. [CrossRef]
66. Wang, L.S.; Shi, W.; Joyce, M.G.; Modjarrad, K.; Zhang, Y.; Leung, K.; Lees, C.R.; Zhou, T.; Yassine, H.M.; Kanekiyo, M.; et al. Evaluation of candidate vaccine approaches for MERS-CoV. *Nat. Commun.* **2015**, *6*, 7712. [CrossRef]
67. Milane, L.; Amiji, M. Clinical approval of nanotechnology-based SARS-CoV-2 mRNA vaccines: Impact on translational nanomedicine. *Drug. Deliv. Transl. Res.* **2021**, *29*, 1–7.
68. Dhama, K.; Sharun, K.; Tiwari, R.; Dadar, M.; Malik, Y.S.; Singh, K.P.; Chaicumpa, W. COVID-19, an emerging coronavirus infection: Advances and prospects in designing and developing vaccines, immunotherapeutics, and therapeutics. *Hum. Vaccines Immunother.* **2020**, *16*, 1232–1238. [CrossRef]
69. Marston, H.D.; Paules, C.I.; Fauci, A.S. Monoclonal Antibodies for Emerging Infectious Diseases-Borrowing from History. *N. Engl. J. Med.* **2018**, *378*, 1469–1472. [CrossRef]
70. Rab, S.; Afjal, A.; Javaid, M.; Haleem, A.; Vaishya, R. An update on the global vaccine development for coronavirus. *Diabetes Metab. Syndr.* **2020**, *14*, 2053–2055. [CrossRef]
71. Chen, Y.N.; Hsueh, Y.H.; Hsieh, C.T.; Tzou, D.Y.; Chang, P.L. Antiviral Activity of Graphene–Silver Nanocomposites against Non-Enveloped and Enveloped Viruses. *Int. J. Environ. Res. Public Health* **2016**, *13*, 430. [CrossRef]
72. Florindo, H.F.; Kleiner, R.; Vaskovich-Koubi, D.; Acúrcio, R.C.; Carreira, B.; Yeini, E.; Tiram, G.; Liubomirski, Y.; Satchi-Fainaro, R. Immune-mediated approaches against COVID-19. *Nat. Nanotechnol.* **2020**, *15*, 630–645. [CrossRef]
73. Shah, V.K.; Firmal, P.; Alam, A.; Ganguly, D.; Chattopadhyay, S. Overview of Immune Response During SARS-CoV-2 Infection: Lessons From the Past. *Front. Immunol.* **2020**, *11*, 1949. [CrossRef]
74. Fleischmann, R.; Genovese, M.C.; Lin, Y.; St John, G.; Van der Heijde, D.; Wang, S.; Gomez-Reino, J.J.; Maldonado-Cocco, J.A.; Stanislav, M.; Kivitz, A.J.; et al. Long-term safety of sarilumab in rheumatoid arthritis: An integrated analysis with up to 7 years' follow-up. *Rheumatology* **2020**, *59*, 292–302. [CrossRef]
75. Chan, W.C.W. Nano research for COVID-19. *ACS Nano.* **2020**, *14*, 3719–3720. [CrossRef]
76. Polyak, B.; Cordovez, B. How can we predict behavior of nanoparticles in vivo? *Nanomedicine* **2016**, *11*, 189–192. [CrossRef]

77. Beyth, N.; Houri-Haddad, Y.; Domb, A.; Khan, W.; Hazan, R. Alternative antimicrobial approach: Nano-antimicrobial materials. *Evid. Based Complement. Altern. Med.* **2015**, *1*, 246012. [CrossRef]
78. Ventola, C.L. Progress in nanomedicine: Approved and investigational nanodrugs. *Pharm. Ther.* **2017**, *42*, 742–755.
79. Caster, J.M.; Patel, A.N.; Zhang, T.; Wang, A. Investigational nanomedicines in 2016: A review of nanotherapeutics currently undergoing clinical trials. *Wiley Interdiscip. Rev. Nanomed. Nanobiotechnol.* **2017**, *9*, e1416. [CrossRef]
80. Sukhanova, A.; Bozrova, S.; Sokolov, P.; Berestovoy, M.; Karaulov, A.; Nabiev, I. Dependence of Nanoparticle Toxicity on Their Physical and Chemical Properties. *Nanoscale Res. Lett.* **2018**, *13*, 44. [CrossRef]

Perspective

A Perspective on Nanotechnology and COVID-19 Vaccine Research and Production in South Africa

Admire Dube [1,*], Samuel Egieyeh [1] and Mohammed Balogun [2,*]

1 School of Pharmacy, University of the Western Cape, Bellville 7535, South Africa; segieyeh@uwc.ac.za
2 Biopolymer Modification and Therapeutics Laboratory, Chemicals Cluster, Council for Scientific and Industrial Research, Brummeria, Pretoria 0001, South Africa
* Correspondence: adube@uwc.ac.za (A.D.); mbalogun@csir.co.za (M.B.)

Abstract: Advances in nanotechnology have enabled the development of a new generation of vaccines, which are playing a critical role in the global control of the COVID-19 pandemic and the return to normalcy. Vaccine development has been conducted, by and large, by countries in the global north. South Africa, as a major emerging economy, has made extensive investments in nanotechnology and bioinformatics and has the expertise and resources in vaccine development and manufacturing. This has been built at a national level through decades of investment. In this perspective article, we provide a synopsis of the investments made in nanotechnology and highlight how these could support innovation, research, and development for vaccines for this disease. We also discuss the application of bioinformatics tools to support rapid and cost-effective vaccine development and make recommendations for future research and development in this area to support future health challenges.

Keywords: COVID-19 and nanotechnology; nanomedicine in South Africa; bioinformatics and vaccine development; vaccine development in South Africa

Citation: Dube, A.; Egieyeh, S.; Balogun, M. A Perspective on Nanotechnology and COVID-19 Vaccine Research and Production in South Africa. *Viruses* **2021**, *13*, 2095. https://doi.org/10.3390/v13102095

Academic Editors: Burtram C. Fielding and Georgia Schäfer

Received: 14 September 2021
Accepted: 12 October 2021
Published: 18 October 2021

1. Introduction

Within 18 months of a pandemic being declared, there were over 220 million confirmed COVID-19 infections and more than 4.6 million deaths globally [1]. Within the same time period, South Africa recorded over 2.8 million cases and over 85,000 deaths [2], the highest on the African continent. The causative organism of COVID-19 is the Severe Acute Respiratory Syndrome Coronavirus 2 (SARS-CoV-2). Coronaviruses are members of the subfamily *Coronavirinae* (family: *Coronaviridae*) that consists of four generations—alphacoronavirus, betacoronavirus, gammacoronavirus and deltacoronavirus—of which, only the betacoronaviruses SARS-CoV and MERS-CoV were regarded as being highly pathogenic prior to 2019 [3]. Acute respiratory distress syndrome as well as extrapulmonary manifestations have been the causes of death [4,5].

An unprecedented effort to develop vaccines to prevent COVID-19 infections and make them clinically available for mass immunizations globally started immediately once the causative organism was identified and its molecular signatures were elucidated in early 2020 [6,7]. The advanced drug delivery systems of nanoparticles have been applied to deliver nucleic acid-based vaccines to cellular and subcellular sites [8]. The research and development efforts for vaccines have been led, as they have traditionally been, largely by countries in the global north [9]. South Africa has emerged as an important global player with a membership in the community of five major emerging economies—which includes Brazil, Russia, India, and China—and is the only African member of the G20, which groups twenty countries that collectively account for 90% of the world's economy. As an "emerging economy" country, it has invested tremendously in scientific research, development, and innovation (RDI) since the dawn of the 21st century. Here, we provide a synopsis of the investments made in nanotechnology by the South African government over the past 15 years as well as their RDI capabilities and highlight how these could support vaccine development for this disease. We also discuss the application of bioinformatics tools to support rapid and cost-effective vaccine development and make recommendations for future research and development in this area that will support preparedness for future health challenges. It is hoped that this perspective article will stimulate more collaborative innovation and product development aimed toward the local production of nanotechnology-based vaccines, therapeutics, and diagnostics for the country and the African continent.

2. Nanotechnology and Vaccines for COVID-19

Some of the vaccine formulations for COVID-19 that are currently used are nanotechnology-based. The mRNA in the Pfizer/BioNTech and Moderna vaccines are delivered in lipid nanoparticle formulations [6,8,10]. In this context, nanotechnology refers to the engineering of particles, which are in the nanoscale (1–1000 nm) and serve as systems to deliver a nucleic acid payload into cellular and subcellular sites in the body. Several reviews have been published, which highlight the utility of nanoparticles in protecting and enhancing the delivery of the mRNA or other antigens against SARS-CoV-2, and the reader is referred to these articles [11–14].

South Africa has yet to develop a COVID-19 vaccine, although researchers in the country also joined the global effort to discover an effective protective agent after the outbreak of the pandemic. The country has been recognized, along with Argentina, as a leader in molecular farming for the development of veterinary therapeutics and vaccines, given the significance of the local livestock industry [15]. The attenuation of a South African lumpy skin disease virus through the gene knockout of a virulent interleukin-10 gene homologue resulted in the demonstration of a vaccine that conferred protection against sheep pox and goat pox [16]. This capability can be conscripted in the development of vaccines against COVID-19.

The know-how to synthesize—and reproducibly manufacture at a large scale—various types of nanoparticles, including lipid nanoparticles, has been around for several decades, and several medicines based on nanotechnology are US Food and Drug Administration (FDA) approved. There are currently over 50 FDA-approved nanomedicines in clinical use [17,18]. Various types of nanoparticles are approved in these medicines, including liposomes, nanocrystals, metallic, polymeric, and lipid nanoparticles [18]. One of the early FDA approvals was in 1995 for a liposomal formulation of the anticancer drug doxorubicin [19]. Onpattro® is an example of a lipid nanoparticle complex containing siRNA that was approved in 2018 [19]. In August 2021, the FDA granted full approval for the Pfizer mRNA lipid nanoparticle COVID-19 vaccine [20]. This endorsement is significant in many respects; not least is that the approved vaccine was developed less than a year after the emergence of the disease from a novel virus. It is also a strong signal of the potential benefits of investments in nanotechnology for modern health.

Nanotechnology has brought several therapeutics to patients over the years. Such medicines may not have made it to market due to unfavorable 'drug-like' properties, or have been repurposed and have had their safety profile and efficacy enhanced [21]. The application of nanotechnology in disease prevention and treatment is acknowledged globally [22,23]. In South Africa, extensive investment in the RDI landscape of nanotechnology formally began in 2005, with the launch of the National Nanotechnology Strategy [24]. Since this time, nanotechnology postgraduate training programs, research and development infrastructure, publications, patents, products, and companies have emanated [24,25]. National nanotechnology characterization centers include those at the Council for Scientific and Industrial Research (https://www.csirnano.co.za/; accessed on 1 October 2021) and the Department of Science and Innovation nanotechnology innovation centers for sensors and bio-labeling, located at Rhodes University and the University of the Western Cape, respectively [24].

3. Potential for Local Research and Development of Nanoparticle-Based COVID-19 Vaccines

A historically high burden of infectious diseases has meant that South African academic and RDI institutions, globally reputed for the quality of their outputs, have had decades of experience in offering cutting-edge solutions in this area. The country has the highest population of HIV-positive people, which has spurred research into nanoparticle-based antivirals. CAP256 is a potent anti-HIV antibody that was isolated from an HIV patient in South Africa in collaboration with international researchers [26]. This antibody has potential therapeutic and preventative properties. Further, CAPRISA 004, a topically applied anti-HIV microbicide was developed and underwent clinical testing in South Africa [27]. These experiences underscore the potential of South African scientists to respond to the need for new antivirals to fight SARS-CoV-2.

South African scientists discovered a new variant of SARS-CoV-2 (beta variant, also known as lineage B.1.351 with E484K and K417N mutations) that was subsequently responsible for the second wave of the disease in late 2020 to early 2021 [28]. The authors also demonstrated that plasma from patients who recovered from infection with the beta variant of SARS-CoV-2 could effectively neutralize the variant of the virus predominantly responsible for the initial wave of COVID-19 in 2020. However, the reverse was not true, as plasma from patients infected during the first wave was largely ineffective against the beta variant. Researchers in South Africa have also made contributions toward the understanding of the pathogenicity of coronavirus envelope proteins [29]. This makes the country the only one on the continent of Africa with the advanced technological capability and expertise to establish COVID-19 vaccine development and manufacture.

Nanoparticles for disease therapeutics have been investigated against infectious diseases, such as HIV/AIDS, malaria, tuberculosis, and cancer [30–34]. The application of 'green nanotechnology', synthesizing therapeutic nanoparticles using extracts of selected plants from the rich land and marine biodiversity of South Africa, has also been achieved. Plants, including *Sutherlandia frutescens*, *Cotyledon orbiculata*, and seaweed extracts of *Codium capitatum*, have been used to synthesize nanoparticles [35–37]. In July 2021, it was announced that Pfizer and BioNTech will—in collaboration with South Africa's Biovac Institute, which already engages in vaccine manufacturing (https://www.biovac.co.za/; accessed on 1 October 2021)—begin production, in 2022, of around 100 million doses per year of their COVID-19 mRNA vaccine for the African continent [38]. Although this does not include the upstream synthesis of the mRNA, it does serve as an acknowledgment of the advancement and potential that exist in the country. A local pharmaceutical company, Aspen Pharmacare, is already in the downstream production of Johnson and Johnson's viral vector COVID-19 vaccine [39]. The application of nanotechnology could also be extended to the delivery of protein-antigen-based vaccines. In the past, researchers at Stellenbosch University have reported the encapsulation of a bacteriocin, Plantaracin 423 (produced by *Lactobacillus plantarum* 423), in nanofibers, which was the first such report [40]. South Africa, therefore, has a great opportunity to innovate in this space and engage in the research, development, and manufacture of its own types of nanoparticles, mRNA, or protein-antigen vaccines for COVID-19. Such technology could also be applied to other diseases endemic in the country.

4. Bioinformatics Approaches to Optimize Research and Development

Bioinformatics pipelines are critical for predicting possible biomolecules that contribute to the understanding of infectious disease mechanisms, therapy, and prevention. Bioinformatics tools and approaches for analyzing biological data generated by genomes, transcriptomics, proteomics, and structural omics are gaining traction for the design of vaccines. These bioinformatics pipelines have been used in South African institutions (e.g., the South African National Bioinformatics Institute) to analyze the sequence of the SARS-CoV-2 genome from the South African population in order to provide a genetic "fingerprint" that can improve understanding and contain the spread of COVID-19 [41,42].

Recently, South African researchers could implement bioinformatics strategies of reverse vaccinology and immuno-informatics in vaccine design and development on a new big data platform for data-intensive research in bioinformatics (http://www.ilifu.ac.za/il/home; accessed on 1 October 2021).

Reverse vaccinology is a way of using bioinformatics tools to identify features in bacteria, viruses, parasites, cancer cells, or allergens that could trigger an immune response [43]. Reverse vaccinology starts with the pathogen's genome and uses it to predict epitopes (antigens) that are most likely to be vaccine candidates. This method decreases the number of proteins to be analyzed, making the selection process easier. Through this process, one can identify antigens present in minute amounts or expressed only at specific phases, and it allows for the study of non-cultivable or dangerous microbes [44]. Immuno-informatics, a sub-discipline of bioinformatics concerned with the computational analysis of immunological data, is also a potent computational tool for vaccine development. Immuno-informatics can minimize the time and cost of vaccine development by predicting optimal antigens, epitopes, carriers, and adjuvants for vaccination. Vaccines against the Ebola virus, HIV-1, Herpes Simplex Virus (HSV)-1 and 2, human norovirus, and *Staphylococcus aureus* are examples that are under development and are using an immuno-informatics approach [45].

Overall, whole-genome sequencing projects and the rise of bioinformatics have triggered the birth of a new era of vaccine research and development, leading to a new generation of vaccines designed by deciphering the information provided by genome sequences and using it to better understand the host–pathogen interactions [46]. Riding on the capacity developed, as well as the gains from the use of bioinformatics to develop insights into the dynamics of other infectious diseases, means that South African scientists are poised to continue applying -omics informatics approaches to advance the prevention and treatment of diseases, including COVID-19 [46,47].

5. Conclusions and Future Perspectives

The emergence of the COVID-19 pandemic was accompanied by a global panic that resulted in an unprecedented shut down of nearly all activities, including most RDI, around the world. With nations having had time to catch their breath, attention is now being turned to how to meet the challenge posed by the novel coronavirus. South Africa is one of only a few African countries that has made significant investments in capabilities that offer leverage to combat COVID-19 and other emerging health threats. As the country regroups from the initial chaos of the pandemic, we have highlighted in this perspective piece the technological and academic resources available to the country from years of RDI investments. Leveraging these capabilities could also light a path out of the enormous economic toll the total and partial lockdowns have had on the country and the cost of procuring vaccines for the entire population. The expertise and resources exist in drug development and vaccine manufacturing: biotechnology, nanotechnology, and bioinformatics are all necessary components for a nanotechnology-based vaccine against COVID-19. Focused collaborations between scientists working in these fields—together with clinicians—are required to drive rapid, cost-effective RDI of vaccines for COVID-19. Such teams could be assembled with the support and funding from the government to research, test, and manufacture these vaccines in a local version of the US government's Operation Warp Speed [48]. Local production has been found to be critical in ensuring adequate supplies within countries. It is without a doubt that African countries should have the capability to manufacture their own vaccines. In April 2021, the Africa Centers for Disease Control launched the Partnerships for African Vaccine Manufacturing (PAVM), having also recognized the need for local manufacturing [49]. Leveraging the vast experience in antiviral research and vaccine development for diseases, such as HIV/AIDS—together with the expertise in nanoparticle synthesis and characterization and bioinformatics—could lead to the rapid development of new vaccines and therapies. Medicines' regulatory support during the vaccine development process is also recognized as a critical component. The capacity to review and grant marketing authorization for vaccines has been demonstrated; however, continued capacity building in this area is required [50]. The soon-to-be-established African Medicines Agency may also play a role in supporting the review of new vaccines and authorization at a continental level.

Author Contributions: Conceptualization, A.D. and M.B.; writing—original draft preparation, review and editing, A.D., S.E. and M.B. All authors have read and agreed to the published version of the manuscript.

Funding: This research was funded by an award from the Fogarty International Center of the National Institutes of Health under Award Number K43TW010371, awarded to A.D., and funding from the National Research Foundation of South Africa (grant ID 109059), awarded to A.D.

Institutional Review Board Statement: Not applicable.

Informed Consent Statement: Not applicable.

Data Availability Statement: Not applicable.

Conflicts of Interest: The authors declare no conflict of interest.

References

1. World Health Organization. WHO Coronavirus (COVID-19) Dashboard. 2021. Available online: https://covid19.who.int/ (accessed on 13 September 2021).
2. Department Health Republic of South Africa. COVID-19 Online Resource & News Portal SAcoronavirus.co.za. 2021. Available online: https://sacoronavirus.co.za/2021/09/13/update-on-covid-19-monday-13-september/ (accessed on 13 September 2021).
3. Cui, J.; Li, F.; Shi, Z.-L. Origin and evolution of pathogenic coronaviruses. *Nat. Rev. Microbiol.* **2019**, *17*, 181–192. [CrossRef] [PubMed]
4. Gupta, A.; Madhavan, M.V.; Sehgal, K.; Nair, N.; Mahajan, S.; Sehrawat, T.S.; Bikdeli, B.; Ahluwalia, N.; Ausiello, J.C.; Wan, E.Y.; et al. Extrapulmonary manifestations of COVID-19. *Nat. Med.* **2020**, *26*, 1017–1032. [CrossRef] [PubMed]
5. Elezkurtaj, S.; Greuel, S.; Ihlow, J.; Michaelis, E.G.; Bischoff, P.; Kunze, C.A.; Sinn, B.V.; Gerhold, M.; Hauptmann, K.; Ingold-Heppner, B.; et al. Causes of death and comorbidities in hospitalized patients with COVID-19. *Sci. Rep.* **2021**, *11*, 4263. [CrossRef] [PubMed]
6. Costanzo, M.; de Giglio, M.A.R.; Roviello, G.N. Anti-Coronavirus Vaccines: Past Investigations on SARS-CoV-1 and MERS-CoV, the Approved Vaccines from BioNTech/Pfizer, Moderna, Oxford/AstraZeneca and others under Development against SARS-CoV-2 Infection. *Curr. Med. Chem.* **2021**, *28*, 1–13. [CrossRef]
7. Tregoning, J.S.; Flight, K.E.; Higham, S.L.; Wang, Z.; Pierce, B.F. Progress of the COVID-19 vaccine effort: Viruses, vaccines and variants versus efficacy, effectiveness and escape. *Nat. Rev. Immunol.* **2021**, *21*, 626–636. [CrossRef]
8. Swingle, K.L.; Hamilton, A.G.; Mitchell, M.J. Lipid Nanoparticle-Mediated Delivery of mRNA Therapeutics and Vaccines. *Trends Mol. Med.* **2021**, *27*, 616–617. [CrossRef]
9. Li, Y.; Tenchov, R.; Smoot, J.; Liu, C.; Watkins, S.; Zhou, Q. A Comprehensive Review of the Global Efforts on COVID-19 Vaccine Development. *ACS Cent. Sci.* **2021**, *7*, 512–533. [CrossRef]
10. Shin, M.D.; Shukla, S.; Chung, Y.H.; Beiss, V.; Chan, S.K.; Ortega-Rivera, O.A.; Wirth, D.M.; Chen, A.; Sack, M.; Pokorski, J.K.; et al. COVID-19 vaccine development and a potential nanomaterial path forward. *Nat. Nanotechnol.* **2020**, *15*, 646–655. [CrossRef]
11. Khurana, A.; Allawadhi, P.; Khurana, I.; Allwadhi, S.; Weiskirchen, R.; Banothu, A.K.; Chhabra, D.; Joshi, K.; Bharani, K.K. Role of nanotechnology behind the success of mRNA vaccines for COVID-19. *Nano Today* **2021**, *38*, 101142. [CrossRef]
12. Buschmann, M.; Carrasco, M.; Alishetty, S.; Paige, M.; Alameh, M.; Weissman, D. Nanomaterial Delivery Systems for mRNA Vaccines. *Vaccines* **2021**, *9*, 65. [CrossRef]
13. Chaudhary, N.; Weissman, D.; Whitehead, K.A. mRNA vaccines for infectious diseases: Principles, delivery and clinical translation. *Nat. Rev. Drug Discov.* **2021**, 1–22. [CrossRef]
14. Hou, X.; Zaks, T.; Langer, R.; Dong, Y. Lipid nanoparticles for mRNA delivery. *Nat. Rev. Mater.* **2021**, 1–17. [CrossRef]
15. Rybicki, E.; Hitzeroth, I.; Meyers, A.; Santos, M.; Wigdorovitz, A. Developing country applications of molecular farming: Case studies in South Africa and Argentina. *Curr. Pharm. Des.* **2013**, *19*, 5612–5621. [CrossRef]
16. Boshra, H.; Truong, T.; Nfon, C.; Bowden, T.R.; Gerdts, V.; Tikoo, S.; Babiuk, L.A.; Kara, P.; Mather, A.; Wallace, D.B.; et al. A lumpy skin disease virus deficient of an IL-10 gene homologue provides protective immunity against virulent capripoxvirus challenge in sheep and goats. *Antivir. Res.* **2015**, *123*, 39–49. [CrossRef]
17. Anselmo, A.C.; Mitragotri, S. Nanoparticles in the clinic: An update post COVID-19 vaccines. *Bioeng. Transl. Med.* **2021**, *6*, e10246. [CrossRef]
18. Bobo, D.; Robinson, K.J.; Islam, J.; Thurecht, K.J.; Corrie, S.R. Nanoparticle-Based Medicines: A Review of FDA-Approved Materials and Clinical Trials to Date. *Pharm. Res.* **2016**, *33*, 2373–2387. [CrossRef]
19. Batty, C.J.; Bachelder, E.M.; Ainslie, K.M. Historical Perspective of Clinical Nano and Microparticle Formulations for Delivery of Therapeutics. *Trends Mol. Med.* **2021**, *27*, 516–519. [CrossRef]

20. Tanne, J.H. COVID-19: FDA approves Pfizer-BioNTech vaccine in record time. *BMJ* **2021**, *374*, n2096. [CrossRef]
21. Dube, A.; Semete-Makokotlela, B.; Ramalapa, B.E.; Reynolds, J.; Boury, F. *Nanomedicines for the Treatment of Infectious Diseases: Formulation, Delivery and Commercialization Aspects*; Routledge: London, UK, 2021.
22. Chang, E.H.; Harford, J.B.; Eaton, M.A.; Boisseau, P.M.; Dube, A.; Hayeshi, R.; Swai, H.; Lee, D.S. Nanomedicine: Past, present and future—A global perspective. *Biochem. Biophys. Res. Commun.* **2015**, *468*, 511–517. [CrossRef]
23. Dube, A. Nanomedicines for Infectious Diseases. *Pharm Res.* **2019**, *36*, 63. [CrossRef]
24. Zhou, D.T.; Maponga, C.C.; Madhombiro, M.; Dube, A.; Mano, R.; Nyamhunga, A.; Machingura, I.; Manasa, J.; Hakim, J.; Chirenje, Z.M.; et al. Mentored postdoctoral training in Zimbabwe: A report on a successful collaborative effort. *J. Public Health Afr.* **2019**, *10*, 1081. [CrossRef]
25. Masara, B.; van der Poll, J.A.; Maaza, M. A nanotechnology-foresight perspective of South Africa. *J. Nanopart. Res.* **2021**, *23*, 92. [CrossRef]
26. Doria-Rose, N.A.; Bhiman, J.N.; Roark, R.S.; Schramm, C.A.; Gorman, J.; Chuang, G.-Y.; Pancera, M.; Cale, E.M.; Ernandes, M.J.; Louder, M.K.; et al. New Member of the V1V2-Directed CAP256-VRC26 Lineage That Shows Increased Breadth and Exceptional Potency. *J. Virol.* **2015**, *90*, 76–91. [CrossRef]
27. Mansoor, L.E.; Karim, Q.A.; Yende-Zuma, N.; MacQueen, K.M.; Baxter, C.; Madlala, B.T.; Grobler, A.; Karim, S.S.A. Adherence in the CAPRISA 004 tenofovir gel microbicide trial. *AIDS Behav.* **2014**, *18*, 811–819. [CrossRef]
28. Tegally, H.; Wilkinson, E.; Giovanetti, M.; Iranzadeh, A.; Fonseca, V.; Giandhari, J.; Doolabh, D.; Pillay, S.; San, E.J.; Msomi, N.; et al. Detection of a SARS-CoV-2 variant of concern in South Africa. *Nature* **2021**, *592*, 438–443. [CrossRef]
29. Schoeman, D.; Cloete, R.; Fielding, B.C. Comparative studies of the seven human coronavirus envelope proteins using topology prediction and molecular modelling to understand their pathogenicity. *bioRxiv* **2021**. [CrossRef]
30. Tshweu, L.L.; Shemis, M.A.; Abdelghany, A.; Gouda, A.; Pilcher, L.A.; Sibuyi, N.R.S.; Meyer, M.; Dube, A.; Balogun, M.O. Synthesis, physicochemical characterization, toxicity and efficacy of a PEG conjugate and a hybrid PEG conjugate nanoparticle formulation of the antibiotic moxifloxacin. *RSC Adv.* **2020**, *10*, 19770–19780. [CrossRef]
31. Melariri, P.; Kalombo, L.; Nkuna, P.; Dube, A.; Hayeshi, R.; Ogutu, B.; Gibhard, L.; Dekock, C.; Smith, P.; Wiesner, L.; et al. Oral lipid-based nanoformulation of tafenoquine enhanced bioavailability and blood stage antimalarial efficacy and led to a reduction in human red blood cell loss in mice. *Int. J. Nanomed.* **2015**, *10*, 1493–1503. [CrossRef]
32. Tshweu, L.; Katata, L.; Kalombo, L.; Chiappetta, D.A.; Hocht, C.; Sosnik, A.; Swai, H. Enhanced oral bioavailability of the antiretroviral efavirenz encapsulated in poly (epsilon-caprolactone) nanoparticles by a spray-drying method. *Nanomedicine* **2014**, *9*, 1821–1833. [CrossRef]
33. Freidus, L.G.; Kumar, P.; Marimuthu, T.; Pradeep, P.; Choonara, Y.E. Theranostic Mesoporous Silica Nanoparticles Loaded with a Curcumin-Naphthoquinone Conjugate for Potential Cancer Intervention. *Front. Mol. Biosci.* **2021**, *8*, 670792. [CrossRef]
34. D'Souza, S.; Du Plessis, S.; Egieyeh, S.; Bekale, R.; Maphasa, R.; Irabin, A.; Sampson, S.; Dube, A. Physicochemical and biological evaluation of curdlan-poly(lactic-co-glycolic acid) nanoparticles as a host-directed therapy against *Mycobacterium tuberculosis*. *J. Pharm. Sci.* **2021**, in press. [CrossRef] [PubMed]
35. Tyavambiza, C.; Elbagory, A.; Madiehe, A.; Meyer, M.; Meyer, S. The Antimicrobial and Anti-Inflammatory Effects of Silver Nanoparticles Synthesised from *Cotyledon orbiculata* Aqueous Extract. *Nanomaterials* **2021**, *11*, 1343. [CrossRef] [PubMed]
36. Dube, P.; Meyer, S.; Madiehe, A.; Meyer, M. Antibacterial activity of biogenic silver and gold nanoparticles synthesized from Salvia africana-lutea and *Sutherlandia frutescens*. *Nanotechnology* **2020**, *31*, 505607. [CrossRef] [PubMed]
37. Kannan, R.; Stirk, W.; Van Staden, J. Synthesis of silver nanoparticles using the seaweed Codium capitatum P.C. Silva (Chlorophyceae). *S. Afr. J. Bot.* **2013**, *86*, 1–4. [CrossRef]
38. Council on Foreign Relations. South Africa's Biovac Strikes Deal to Make COVID-19 Vaccine. 2021. Available online: https://www.cfr.org/blog/south-africas-biovac-strikes-deal-make-covid-19-vaccine (accessed on 13 September 2021).
39. News, A.H. Aspen Confirms Release of COVID-19 Vaccines to Johnson & Johnson for Supply to South Africa. 2021. Available online: https://www.aspenpharma.com/2021/07/26/aspen-confirms-release-of-covid-19-vaccines-to-johnson-johnson-for-supply-to-south-africa/ (accessed on 13 September 2021).
40. Heunis, T.D.J.; Botes, M.; Dicks, L.M.T. Encapsulation of *Lactobacillus plantarum* 423 and its Bacteriocin in Nanofibers. *Probiotics Antimicrob. Proteins* **2010**, *2*, 46–51. [CrossRef] [PubMed]
41. Mulder, N.J.; Christoffels, A.; De Oliveira, T.; Gamieldien, J.; Hazelhurst, S.; Joubert, F.; Kumuthini, J.; Pillay, C.S.; Snoep, J.L.; Bishop, O.T.; et al. The Development of Computational Biology in South Africa: Successes Achieved and Lessons Learnt. *PLoS Comput. Biol.* **2016**, *12*, e1004395. [CrossRef] [PubMed]
42. Allam, M.; Ismail, A.; Khumalo, Z.T.H.; Kwenda, S.; van Heusden, P.; Cloete, R.; Wibmer, C.K.; Mtshali, P.S.; Mnyameni, F.; Mohale, T.; et al. Genome Sequencing of a Severe Acute Respiratory Syndrome Coronavirus 2 Isolate Obtained from a South African Patient with Coronavirus Disease 2019. *Microbiol. Resour. Announc.* **2020**, *9*, e00572-20. [CrossRef]
43. Enayatkhani, M.; Hasaniazad, M.; Faezi, S.; Gouklani, H.; Davoodian, P.; Ahmadi, N.; Einakian, M.A.; Karmostaji, A.; Ahmadi, K. Reverse vaccinology approach to design a novel multi-epitope vaccine candidate against COVID-19: An in silico study. *J. Biomol. Struct. Dyn.* **2021**, *39*, 2857–2872. [CrossRef]
44. Soltan, M.A.; Magdy, D.; Solyman, S.M.; Hanora, A. Design of Staphylococcus aureus New Vaccine Candidates with B and T Cell Epitope Mapping, Reverse Vaccinology, and Immunoinformatics. *Omics* **2020**, *24*, 195–204. [CrossRef]

45. Oli, A.N.; Obialor, W.O.; Ifeanyichukwu, M.O.; Odimegwu, D.C.; Okoyeh, J.N.; Emechebe, G.O.; Adejumo, S.A.; Ibeanu, G.C. Immunoinformatics and Vaccine Development: An Overview. *ImmunoTargets Ther.* **2020**, *9*, 13–30. [CrossRef]
46. Egieyeh, S.; Egieyeh, E.; Malan, S.; Christofells, A.; Fielding, B. Computational drug repurposing strategy predicted peptide-based drugs that can potentially inhibit the interaction of SARS-CoV-2 spike protein with its target (humanACE2). *PLoS ONE* **2021**, *16*, e0245258. [CrossRef]
47. Chukwudozie, O.S.; Gray, C.M.; Fagbayi, T.A.; Chukwuanukwu, R.C.; Oyebanji, V.O.; Bankole, T.T.; Adewole, R.A.; Daniel, E.M. Immuno-informatics design of a multimeric epitope peptide based vaccine targeting SARS-CoV-2 spike glycoprotein. *PLoS ONE* **2021**, *16*, e0248061. [CrossRef]
48. Kim, J.H.; Hotez, P.; Batista, C.; Ergonul, O.; Figueroa, J.P.; Gilbert, S.; Gursel, M.; Hassanain, M.; Kang, G.; Lall, B.; et al. Operation Warp Speed: Implications for global vaccine security. *Lancet Glob. Health* **2021**, *9*, e1017–e1021. [CrossRef]
49. Irwin, A. How COVID spurred Africa to plot a vaccines revolution. *Nature* **2021**. online ahead of print. [CrossRef]
50. Semete-Makokotlela, B.; Mahlangu, G.N.; Mukanga, D.; Darko, D.M.; Stonier, P.; Gwaza, L.; Nkambule, P.; Matsoso, P.; Lehnert, R.; Rosenkranz, B.; et al. Needs-driven talent and competency development for the next generation of regulatory scientists in Africa. *Br. J. Clin. Pharmacol.* **2021**, 1–8. [CrossRef]

Article

Rapid and Successful Implementation of a COVID-19 Convalescent Plasma Programme—The South African Experience

Tanya Nadia Glatt [1,*], Caroline Hilton [2], Cynthia Nyoni [1], Avril Swarts [1], Ronel Swanevelder [1], James Cowley [3], Cordelia Mmenu [3], Thandeka Moyo-Gwete [4,5], Penny L. Moore [4,5], Munzhedzi Kutama [3], Jabulisile Jaza [3], Itumeleng Phayane [3], Tinus Brits [6], Johan Koekemoer [6], Ute Jentsch [1], Derrick Nelson [1], Karin van den Berg [1,7,8,†] and Marion Vermeulen [3,8,†]

1 Medical Division, South African National Blood Service, Roodepoort 1709, South Africa; cynthia.nyoni@sanbs.org.za (C.N.); avril.swarts@sanbs.org.za (A.S.); Ronel.Swanevelder@sanbs.org.za (R.S.); Ute.Jentsch@sanbs.org.za (U.J.); Derrick.Nelson@sanbs.org.za (D.N.); Karin.vandenBerg@sanbs.org.za (K.v.d.B.)
2 Medical Division, Western Cape Blood Service, Cape Town 7405, South Africa; Caroline@wcbs.co.za
3 Operations Division, South African National Blood Service, Roodepoort 1709, South Africa; James.Cowley@sansb.org.za (J.C.); Cordelia.Mmenu@sanbs.org.za (C.M.); Munzhedzi.Kutama@sanbs.org.za (M.K.); Jabulisile.Jaza@sanbs.org.za (J.J.); Itumeleng.Phayane@sanbs.org.za (I.P.); Marion.Vermeulen@sanbs.org.za (M.V.)
4 National Institute for Communicable Diseases of the National Health Laboratory Services, Johannesburg 2192, South Africa; thandekam@nicd.ac.za (T.M.-G.); pennym@nicd.ac.za (P.L.M.)
5 MRC Antibody Immunity Research Unit, School of Pathology, Faculty of Health Sciences, University of the Witwatersrand, Johannesburg 2000, South Africa
6 Information Technology Division, South African National Blood Service, Roodepoort 1709, South Africa; Tinus.Brits@sanbs.org.za (T.B.); Johan.Koekemoer@sanbs.org.za (J.K.)
7 Division of Clinical Haematology, Department of Medicine, Faculty of Health Sciences, University of Cape Town, Cape Town 7935, South Africa
8 Division of Clinical Haematology, University of the Free State, Bloemfontein 9301, South Africa
* Correspondence: tanya.glatt@sanbs.org.za
† K.v.d.B. and M.V. contributed equally to the project and the manuscript.

Citation: Glatt, T.N.; Hilton, C.; Nyoni, C.; Swarts, A.; Swanevelder, R.; Cowley, J.; Mmenu, C.; Moyo-Gwete, T.; Moore, P.L.; Kutama, M.; et al. Rapid and Successful Implementation of a COVID-19 Convalescent Plasma Programme—The South African Experience. *Viruses* **2021**, *13*, 2050. https://doi.org/10.3390/v13102050

Academic Editors: Burtram C. Fielding and Georgia Schäfer

Received: 20 August 2021
Accepted: 4 October 2021
Published: 12 October 2021

Abstract: Background: COVID-19 convalescent plasma (CCP) has been considered internationally as a treatment option for COVID-19. CCP refers to plasma collected from donors who have recovered from and made antibodies to SARS-CoV-2. To date, convalescent plasma has not been collected in South Africa. As other investigational therapies and vaccination were not widely accessible, there was an urgent need to implement a CCP manufacture programme to service South Africans. Methods: The South African National Blood Service and the Western Cape Blood Service implemented a CCP programme that included CCP collection, processing, testing and storage. CCP units were tested for SARS-CoV-2 Spike ELISA and neutralising antibodies and routine blood transfusion parameters. CCP units from previously pregnant females were tested for anti-HLA and anti-HNA antibodies. Results: A total of 987 CCP units were collected from 243 donors, with a median of three donations per donor. Half of the CCP units had neutralising antibody titres of >1:160. One CCP unit was positive on the TPHA serology. All CCP units tested for anti-HLA antibodies were positive. Conclusion: Within three months of the first COVID-19 diagnosis in South Africa, a fully operational CCP programme was set up across South Africa. The infrastructure and skills implemented will likely benefit South Africans in this and future pandemics.

Keywords: convalescent plasma; SARS-CoV-2; COVID-19

1. Introduction

SARS-CoV-2, the virus that causes COVID-19, poses a significant threat to global health. The lack of definitive treatment or widely accessible effective prevention has led many to consider COVID-19 convalescent plasma (CCP) as a potential therapeutic option. CCP refers to plasma collected from donors who have recovered from COVID-19 and, therefore likely to have produced neutralising antibodies against SARS-CoV-2 [1]. Convalescent plasma (CP) has been used successfully as a form of passive immunity for previous viral infections, including severe influenza, severe acute respiratory syndrome (SARS)-CoV, Middle East respiratory syndrome (MERS)-CoV and to some extent, Ebola virus disease [2–4]. It was hypothesised that the infusion of plasma with virus-specific antibodies might yield immediate passive immunity to the recipient and improve viral clearance [5].

Source plasma is a procedure whereby a donor's plasma is collected through apheresis techniques, following which their cellular components are returned. Although hyper-immune plasma for Hepatitis B Virus (HBV) and source plasma for intravenous immunoglobulins (IVIG's) from donors are routinely collected by apheresis and produced in South Africa, the country has not previously produced CP. With the outbreak of COVID-19 in South Africa from March 2020, the South African National Blood Service (SANBS) and the Western Cape Blood Service (WCBS) collaborated in setting up a national CCP programme. With a national footprint, facilities and processes for donor recruitment, plasma apheresis, infectious disease testing, component processing and inventory management already in place, blood services are advantageously placed to rapidly implement such CCP programmes. In a few short months, the SANBS and WCBS teams were successful in implementing a CCP programme.

Data is ever-evolving in the domain of COVID-19 treatments, and CCP has been gathering interest around the world as a potential therapeutic option, with international guidelines and publications on its production and use being released and updated continuously [6]. Of particular interest is the viral evolution and formation of new variants and the possible consequences they have on the efficacy of CCP from an alternate variant. Initially, a number of cohort studies [7–9] showed the effectiveness of CCP, requiring the need for large randomised control trials to establish both efficacy and safety. Our programme was intended for use in a phase III randomised controlled trial. Unfortunately, when the results of the large randomised control trial became available, the use of CCP in hospitalised patients with COVID-19 pneumonia showed little or no benefit [10–13]. However, there is evidence of clinical efficacy in specific population groups such as older, at-risk patients [14], early in the disease [15], and patients with immunosuppression secondary to haematology malignancies [16]. Introducing a new blood product programme at the height of a pandemic posed multiple challenges, including regulatory, logistical, ethical and scientific considerations [17], especially in resource-restricted settings [18]. This manuscript describes our efforts in addressing these challenges in South Africa.

2. Materials and Methods

2.1. Study Setting

SANBS and WCBS collect approximately 960,000 and 150,000 donations per annum, respectively, from a population of ~60 million people. SANBS covers 8 of the 9 provinces, and WCBS covers the Western Cape Province. The 8 provinces that SANBS services are structured into 7 collection and processing zones, with donation testing laboratories located in two of the zones. Both blood services are vein-to-vein organisations in that they collect, process, test, store and manage the compatibility testing and issuing processes of blood transfusion.

2.2. Regulatory Approval

Given the lack of empirical evidence on the efficacy of CCP at the start of the epidemic, the CCP programme in South Africa was initially limited to a clinical trial setting, which required multiple layers of regulatory approval. Two independent protocols were developed—one for CCP manufacturing and one for CCP clinical use; they were named the PROTECT-Donor (PROspective, randomised, placebo-controlled, double-blinded, phase

III clinical trial of the Therapeutic use of convalEsCenT plasma in the treatment of patients with moderate to severe COVID-19) and PROTECT-Patient, respectively. The PROTECT-Donor protocol, which included donor recruitment, informed consent, collection, testing, processing and storage of CCP, required approval from the SANBS Scientific Review and Human Research Ethics Committees (2019/0519). The PROTECT-Patient protocol required additional regulatory authority approvals, including the South African Health Products Regulatory Authority (SAHPRA) for the use of an investigational product, Department of Health and site-specific approvals from clinical sites and registration on trials.gov (NCT04516811). The study was conducted according to the guidelines of the Declaration of Helsinki and approved by the Human Research Ethics Committee of the South African National Blood Service (protocol code 2019/0519 on the 21 April 2020).

During the COVID-19 pandemic in South Africa, lockdown and travel restrictions and regulations were instituted by the South African government controlling various activities based on alert levels 1–5 [19]. These restrictions included limited movement outside one's primary residence, closure of all non-essential facilities and bans on the sale of alcohol. These restrictions significantly impacted the ability of the team to obtain these regulatory approvals and implement the programme.

2.3. Study Preparation and Staff Training

SANBS was collecting source plasma as a routine product prior to the start of the study, with an established infrastructure already in place. WCBS had not collected source plasma previously, so the placement of two apheresis machines in their Specialised Donation Headquarter Clinic in Cape Town and full training of three staff members was required. After obtaining the required regulatory approvals, detailed work instructions regarding all aspects of recruitment, collection, testing, processing and transport of CCP for the study were compiled by both blood services along with specific forms to facilitate the transport of the study samples and products to the correct locations. Study-specific training and competency assessments took place at the relevant sites, which was challenging at times due to the travel restrictions. Two nurses were recruited to manage the study and perform recruitment, and a medical technologist was assigned to assist in the management of samples at SANBS to accommodate the study workload.

2.4. Donor Recruitment

Several platforms were utilised to recruit participants, including media coverage through television and radio interviews and advertisements on the SANBS and WCBS websites. The tele-recruiters and customer service staff at both blood collection services were educated about the study and asked to advertise to existing donors. One targeted strategy was the involvement of 'recruitment partners' that included treating doctors and community ambulance organisations who encouraged eligible COVID-19 patients to consider registering as donors after they had recovered.

Participating donors were required to meet the routine criteria for source plasma collection, be between the ages of 18 to 65 years and have a confirmed positive COVID-19 PCR laboratory test. Initially, donors were required to be symptom-free for at least 28 days, although this was later reduced to 14 days to align with updated international guidelines [20]. Due to their higher rates of human leucocyte antigen (HLA)/human nuclear antigen (HNA) antibodies, previously pregnant females were initially excluded from recruitment as such antibodies are associated with a higher risk of antibody-associated transfusion-related acute lung injury (TRALI). However, a protocol amendment was approved to include previously pregnant females conditioned upon a negative HLA/HNA antibody test.

Donor registration was done through an established link on the SANBS website, which guided potential donors through a pre-screening questionnaire where they could record their contact details. The research nurses contacted the registered donors to perform a telephonic interview. Once preliminary eligibility had been confirmed, information regarding the study was provided, and telephonic consent was obtained. An appointment for a pre-screening visit was arranged with the participating donor centre most conveniently situated to the donor.

2.5. Donor Screening and Plasma Collection

The purpose of the initial pre-screening visit was to ensure that the donor met all physical and laboratory criteria for study participation. This involved completion of the routine blood donation questionnaire, testing of their haemoglobin levels by a quantitative point-of-care device and examination of their veins for suitability to donate plasma. Provided these criteria were met, enrolment was completed by the donor signing the informed consent document and having pre-screening blood specimens taken (routine viral and immunohaematology blood donation testing, full blood count and SARS-CoV-2 antibody titre testing). A provisional appointment was made for the donor's first plasma donation. It was explained that the donor needed to have a sufficiently high SARS-CoV-2 binding antibody level (defined as a SARS-CoV-2 spike enzyme-linked immunosorbent assay (ELISA) antibody optical density (OD_{450nm}) value > 0.4) to participate in the study and that this result would be communicated telephonically. If so, they would return to the clinic for their first CCP donation, provided routine donation questionnaire answers, and haemoglobin screening was passed on the day.

CCP donations took place at donor sites across all nine provinces and followed the same collection protocol as for source plasma. Donors were encouraged to donate plasma every two weeks, and follow-up donation dates were ideally agreed upon before the donor left the clinic on the day of donation. Routine source plasma donor screening was performed, and samples for SARS-CoV-2 binding antibody titre levels were taken at each visit. In the event that a donor's antibody titre fell below the required value of SARS-CoV-2 ELISA spike OD_{450nm} < 0.4, they would be offered repeat testing a month later. If this result was persistently low, the donor was thanked for their participation in the study and motivated to become a regular apheresis or whole blood donor. As for other blood donors in South Africa, CCP donors were not remunerated in this study. Communication between CCP donors and the study staff was made by email, telephone and via social media applications such as WhatsApp®.

During all contact visits, infection prevention control (IPC) measures were followed, such as hand hygiene practices, environmental infection control and wearing personal protective equipment (PPE), including masks. All donor centres and staff adhered to the South African guidelines for quarantine and isolation in relation to COVID-19 exposure and infection [21] at all times.

2.6. Donor Testing

Routine blood donation testing, including transfusion transmissible infectious disease testing (TTID), was performed in-house in line with the standard blood screening algorithm using the Beckman Coulter PK7300 instrument (Brea, CA, USA) for blood grouping, ABO titres and syphilis (TPHA) testing; Sitetech Erythra instrument (Grifols, Spain) for irregular antibody screening; Abbott Alinity S instrument (Delkenhein, Germany) for HIV, Hepatitis B virus(HBV) and Hepatitis C virus (HCV serology testing; and Grifols Panther instrument (Barcelona, Spain) for individual-donation nucleic acid testing (ID-NAT) for HIV, HBV and HCV.

The National Institute of Communicable Disease (NICD) performed SARS-CoV-2 Spike ELISA antibody titre testing using an in-house assay based on the assay developed by Mount Sinai [22]. A spike OD_{450nm} of $\geq$0.4 was considered positive, and all donors with values higher than this were recruited for the study. Testing for neutralising antibodies (nAb) was performed by the NICD on stored samples, using a previously validated published method [23]. SARS-CoV-2 Wuhan-1/D614 spike proteins were expressed in Human Embryonic Kidney (HEK) 293F suspension cells by transfecting the cells with the spike plasmid. After incubating for six days at 37 °C, 70% humidity and 10% CO_2, proteins were first purified using a nickel resin followed by size-exclusion chromatography.

Relevant fractions were collected and frozen at −80 °C until use. Two µg/mL of spike protein were used to coat the 96-well high-binding plates and incubated overnight at 4 °C. The plates were incubated in a blocking buffer consisting of 5% skimmed milk powder, 0.05% Tween 20, 1× phosphate-buffered saline. Plasma samples were diluted to a 1:100 starting dilution in a blocking buffer and added to the plates. Secondary antibody was diluted to 1:3000 in blocking buffer and added to the plates followed by TMB substrate (Thermofisher Scientific, Waltham, MA, USA). Upon stopping the reaction with 1 M H_2SO_4, absorbance was measured at a 450 nm wavelength. In all instances, mAb CR3022 was used as a positive control, and palivizumab was used as a negative control. In line with published literature [20] and guidance from researchers in the clinical arm of the CCP trial, CCP was only collected from donors who had nAb titres greater than 1:160 to ensure that the therapeutic dose would be adequate. While nAb testing is believed to be superior, it is a time-consuming and labour-intensive process that requires a Biosafety Level 2 laboratory. The preliminary results of anti-spike binding antibodies and nAb showed a good correlation as in several other studies [24,25]. To ensure efficient use of available resources, a decision was taken to prioritise samples with spike OD_{450nm} values > 2 for nAb testing.

Previously pregnant females were screened for HLA Class I and II and HNA antibodies utilising One Lambda's LabScreen Single Antigen (SA) I and II and LabScreen Multi-kits (Thermo Fisher Scientific, Waltham, MA, USA). These tests were performed on the Luminex 200 and FM3D instruments (Thermo Fisher Scientific, Waltham, MA, USA) as per the manufacturer's instructions. A cut-off mean fluorescence intensity (MFI) value of 2000 was used for SA I and II HLA antibodies. Donors with a mean MFI > 2000 were considered high risk for TRALI [26]. The assay also provided a panel reactive antibody (PRA) result for SA I and II results. This is a common method to determine the level of sensitisation, expressed as a percentage (%). A high PRA% means that the individual is primed to react immunologically against a large proportion of the populations' HLA antigens. The MFI and PRA% were analysed together to assess risk for TRALI.

2.7. Processing and Inventory Management

CCP donations collected by SANBS were transported to the zone processing site where they were blast frozen to −60 °C over 45–60 min, then sent to a centralised site for pathogen reduction treatment (PRT). Units collected by the WCBS were processed, pathogens were reduced on-site at the headquarters facility, and only CCP units that met the antibody criteria underwent PRT. At SANBS, PRT was performed using the Intercept® PRT system (Cerus Corporation, Concord, CA, USA); WCBS used the Mirasol® system (Terumo BCT, Lakewood, CO, USA) and performed the PRT.

Early in the study, a number of damaged CCP units were identified at SANBS during the initial thawing process. The study team worked quickly to investigate and implement changes to minimise CCP unit loss. These investigations included an interrogation of the bag specifications and suitability, removal of excess air at the time of sterility sampling, reducing freezer temperatures and run times to −30 °C and 45 min, respectively, placing bubble wrap between the plasma bag and dry ice in the transport process and seeking advice from the PRT manufacturer. Staff were instructed to take additional care when handling the units, to physically inspect all units, place them directly in bubble wrap bags and limit packaging to a maximum of ten units per crate. A different hamper was introduced for transporting units to the central processing centre to minimise movement during transport. A warm air plasma thawer replaced the water bath, which limited both breakages and potential bacterial contamination.

The pathogen-reduced CCP units were distributed as required from the processing sites at SANBS and WCBS to the different blood banks serving the hospitals participating in the clinical trial. CCP products were managed as per routine inventory procedures in the blood management system, identified by specific product codes. Routine blood ordering request forms were used by the clinical trial staff to order the products, and standard procedures for the issuing of blood products were followed. ABO blood group-specific products were selected for study patients and were thawed in the blood bank prior to issue. As the clinical trial was double-blind and placebo-controlled, the blood bank staff received instruction from the Randomisation Officer to prepare either the CCP or the saline placebo, which were wrapped to ensure the products were near indistinguishable, to issue to the study nurse.

2.8. Information Technology and Analytical Support

An application developed on the K2 cloud-based software (Nintex Group PTY LTD., Bellevue, WA, USA) was used to register and manage potential CCP donors, irrespective of the platform where they were recruited. All donor information, including contact details and information on each donation, were captured on this system. Once a donor arrived for a pre-donation screening, the donor was registered on the blood establishment computer system (BECS), and all subsequent donations, testing and management of processed blood products were captured as per routine core processes. The SANBS and WCBS BECS were modified to include CCP donations as well as the various products made through the processing of the donation. In addition, the BECS systems were modified to include the anti-SARS-CoV-2 test results. These data were included in the routine extraction, transformation and loading processes feeding the SANBS data warehouse. Operational reports combining data from WCBS and SANBS were developed by the Business Intelligence department for the day-to-day management of CCP donors, donations and products.

3. Results

3.1. Regulatory Requirements

The first case of COVID-19 was reported in South Africa on the 5 March 2020. The SANBS Scientific Research Committee and Human Research Ethics Committee approval for the CCP manufacture programme was granted in April 2020, and collections started on the 1 May 2020. The clinical CCP programme began later on the 30 September when SAPHRA approval was obtained. In the period between CCP manufacture approval and CCP clinical-use approval, CCP was collected, tested, manufactured, pathogen-reduced and stored, but not released to patients. The CCP manufacture program continued until March 2021.

3.2. Donor Recruitment and Plasma Collection

3.2.1. Donor Enrolment and Recruitment

A total of 660 people expressed interest in donating CCP; of these, 156 (24%) people could not be contacted due to an incomplete registration, 157 (24%) people were excluded during the telephonic screening process and assessment appointments were scheduled for 347 (53%) people, two of whom did not attend. After routine blood donor screening, physical assessment and blood sampling, a further 104 people were excluded, mainly due to not returning for the CCP donation. A total of 243 donors were accepted for the study (Table 1). Of these, the majority (52.7%) registered on the website, followed by 22.6% making direct telephonic contact via the advertised numbers and 17.7% expressing interest at a blood donation clinic (Figure 1).

Table 1. Donor registrations.

	Number of Donors	%
Registrations on SANBS website	660	100
Incomplete registration	156	23.6
Failed telephonic screening interview	157	23.8
Failed pre-screening assessment at donation site	104	15.8
Enrolled	243	36.8

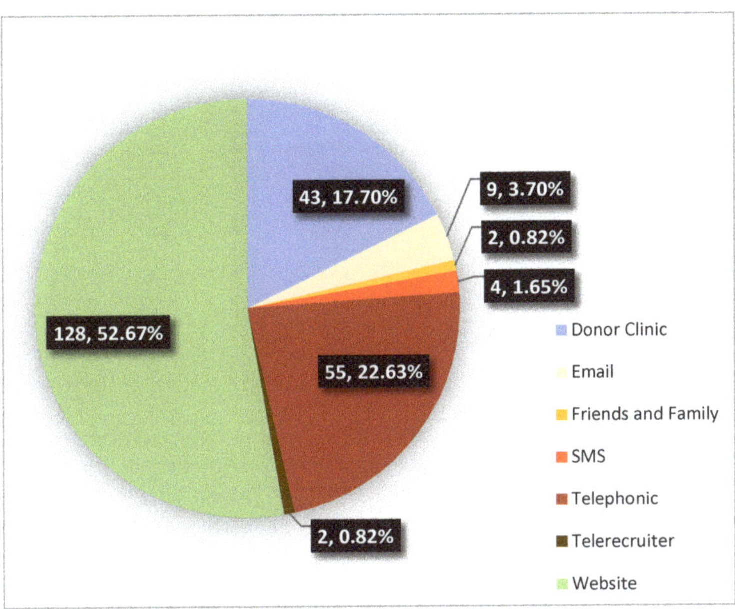

Figure 1. Recruitment platforms used for donor recruitment (*n*; %).

3.2.2. Donor and Donation Data

The age, gender, ethnicity and blood groups of enrolled donors are shown in Table 2. Most donors were aged 20–29 years (*n* = 75; 30.9%), followed by those aged 30–39 years (*n* = 51; 21.0%) and 40–49 years (*n* = 50; 20.6%). Donors were predominantly male (*n* = 151; 62.1%), White (*n* = 178; 73.3%) and belonged to blood groups O (*n* = 102; 42.0%) and A (n = 98; 40.3%).

Table 2. Age group, gender, blood group and ethnicity classification of enrolled donors.

		Number of Donors	%	Number of Donations	%
Total		243	100	987	100
Age Group (years)	18–19	7	2.9	14	1.4
	20–29	75	30.9	267	27.1
	30–39	51	21.0	166	16.8
	40–49	50	20.6	222	22.5
	50–59	45	18.5	244	24.7
	60–69	15	6.2	74	7.6
Gender	Female	92	37.9	349	35.4
	Male	151	62.1	638	64.6
Blood Group	Group A	98	40.3	396	40.1
	Group AB	11	4.5	47	4.8
	Group B	32	13.2	148	15.0
	Group O	102	42.0	396	40.1
Ethnicity	Asian/Indian	20	8.2	92	9.3
	Black African	22	9.1	80	8.1
	Coloured	19	7.8	87	8.8
	Unknown	4	1.7	19	1.9
	White	178	73.3	709	71.8

A total of 987 CCP units were collected from all nine provinces (Table 3). Each CCP unit comprised of approximately 650 mL and was aliquoted, when possible, into two CCP "doses", sufficient for approximately 1974 patients. The highest proportion of donors were recruited from the Egoli zone (35.0%), followed by the Western Cape (16.5%), with the least number of donors originating from the Free State/North Cape zone (1.2%).

Table 3. Location of donors and total collections.

		Number of Donors	%	Number of Donations	%
Zone	Eastern Cape	27	11.1	124	12.6
	Egoli	85	35.0	370	37.5
	Free State/North Cape	3	1.2	13	1.3
	KwaZulu Natal	16	6.6	91	9.2
	Mpumalanga	10	4.1	40	4.1
	Northern	37	15.2	158	16.0
	Vaal	25	10.3	89	9.0
	Western Cape	40	16.5	102	10.3
	Totals	243	100	987	100

The numbers of CCP donations per donor are shown in Table 4. The median number of donations made by a study participant was three, with just over a quarter of donors making only a single donation. The most donations by a study participant was 17.

Table 4. Number of donations per donor.

Number of Donations	Donors	% of Donors
1	65	26.8
2	34	14.0
3	37	15.2
4	25	10.3
5	20	8.2
6	20	8.2
7	6	2.5
8	7	2.9
9	6	2.5
10	6	2.5
11	7	2.9
12	5	2.1
13	1	0.4
14	1	0.4
15	2	0.8
17	1	0.4

3.2.3. Test Results

All donations tested negative for HIV, HBV and HCV by both serological and NAT testing. One donor was excluded from the study based on a positive TPHA result.

Of the 987 CCP donations collected, 940 were tested for binding antibodies by the spike ELISA of which 54 (6%) were classified as sero-silent. In total, 886 CCP units had an $OD_{450nm} \geq 0.4$ with a median (IQR) of 1.9 (1.3–2.6). The CCP units (344) were tested for nAb titres (Figure 2), of which 50% (n = 172) were below the cut-off of 1:160. There were 80 units (23%) with nAb titres between 160–299, 72 units (21%) with titres between 300–999 and 20 units (6%) with titres $\geq$ 1000. There was a good correlation between nAB titres and spike OD (Figure 3).

Anti-HNA antibodies were detected in three (8.8%) serum samples from previously pregnant donors. Anti-HLA antibodies were detected in all 34 serum samples from previously pregnant donors: 25 samples were positive for both HLA Class I and II antibodies, and the remaining nine samples were either positive for HLA Class I or Class II antibodies. The median (range) PRA% and MFI for SA I was 22% (5–52) and 2597 (993-25017), respectively. For SA II, the PRA% was 22% (1–47), and MFI was 3139 (1008-18877). When results of anti-HNA antibody, anti-HLA antibody MFI and PRA% were analysed together, all 34 samples tested were considered as high risk for causing TRALI, and these donors were excluded from donation.

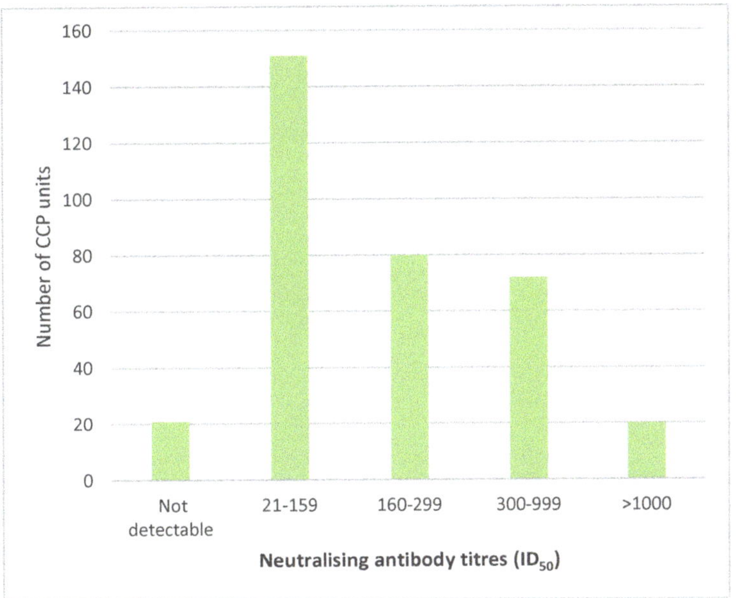

Figure 2. Neutralising antibody titres (ID_{50}) of CCP units tested.

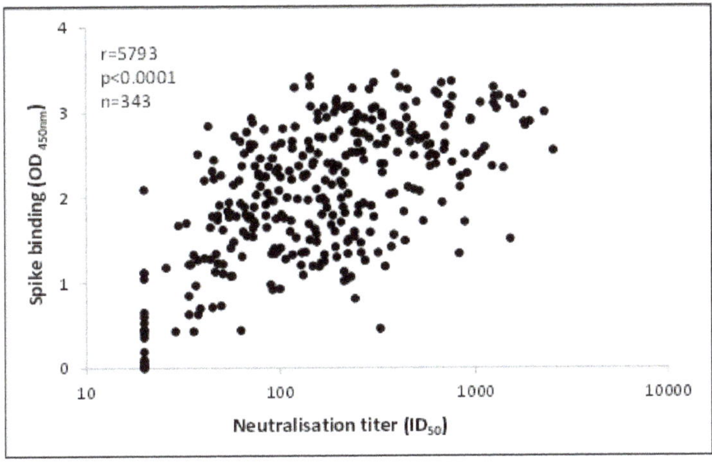

Figure 3. Comparison of neutralizing antibody titers and spike binding antibody responses (Optical Density at 450 nm) of CCP units.

4. Discussion

Prior to this study, CCP had not been produced in South Africa. As a novel product intended for clinical use in an acute pandemic setting, multiple, complex regulatory approvals were required, with delays and hurdles threatening the programme. To compensate for this, the programme was separated into two arms: the CCP manufacture arm (PROTECT-Donor) and the CCP clinical-use arm (PROTECT-Patient), each of which had separate protocols and approvals. Constructing the programme in this manner was time-efficient, as CCP could be manufactured while lengthy and complex regulatory approvals for clinical research use were not yet in place and time-consuming processes to find eligible hospitals with qualified doctors to take part in the clinical trial were being finalised. Within three months of the first confirmed case of COVID-19 in the country, the blood services collected the first CCP donation. Once the clinical trial was set up, the CCP product was readily available, and patient recruitment, at the height of the second wave in South Africa, was not limited by product availability.

The well-established source plasma programme at SANBS allowed for rapid implementation of a national CCP programme. Various new challenges were encountered during different stages of implementation, including logistical (ensuring staff and donor access to donor centres during the national lockdown and travel restrictions and minimising the cannibalising of whole blood donations); PPE and IPC requirements (adequate social distancing and associated decrease in the number of persons at a donation site, resulting in staggering staff shifts and reduced output); human resources (staff concerns regarding recently infected donors and reduced staff pool due to isolation and quarantine); and scientific (restricted access to and turnaround time of SARS-CoV-2 antibody testing and the rapidly evolving international recommendations of CCP titre acceptability criteria). Pathogen reduction technology had not been performed in South Africa prior to the study, and therefore procurement sourcing, installation qualification, validations, standard operating procedures and training had to be rapidly implemented.

Donor recruitment is central to a successful CCP programme. More than half of the donors contacted our research team via the website. The website was easy to navigate and provided detailed information, including the inclusion and exclusion criteria for donating, with the advantage of excluding ineligible donors upfront and negating the need for one-on-one telephonic screening.

The demographical distribution of CCP donors was similar to routine blood donors in some respects, including age and blood group. White donors were overrepresented compared to the active whole blood donor base. However, females were underrepresented; initially, all of the previously pregnant females were excluded, and the youngest age group was not well represented. This may be due to younger adults (under the age of 20) experienced milder symptoms (or asymptomatic disease) and therefore did not present for SARS-CoV-2 testing. In addition, the closure of high schools and universities during the COVID-19 lockdown periods, which are the traditional recruitment centres for young donors, may have contributed to this underrepresentation.

The HLA and HNA antibody test results were informative as they showed that, as expected, previously pregnant females had a high rate of antibody positivity. The presence and strength of these antibodies are postulated to be related to the number of previous pregnancies and time since the most recent pregnancy [27,28]. In view of the high cost and time constraint of HLA and HNA antibody testing, it may be prudent to exclude all previously pregnant females regardless of parity. This is specifically applicable in lower- and middle-income countries with limited resources and availability of specialised tests.

Collaboration and partnership with key stakeholders was a cornerstone of this programme's success. Close collaboration between SANBS and WCBS in all aspects of this study allowed for the inclusion of donors and recipients across South Africa to donate and receive CCP in a near-seamless fashion, ensuring standardised approaches to both donor and patient selection, monitoring and reporting. Partnering with the NICD enabled access to specialised research SARS-CoV-2 antibody testing, and successful grant applications allowed for prompt funding of additional staff, equipment and testing requirements.

This study had some limitations. Data were not collected on which recruitment type was the most successful (such as radio interviews, posters, recruitment by healthcare workers), which would have been helpful for future projects. Although the website was a successful form of donor recruitment, if CCP demand had been significantly increased, the donor pool would have been insufficient to meet demands, and alternate methods of recruitment would have been required.

Of major importance, our CCP programme (PROTECT-Donor) occurred during the first wave of COVID-19 in South Africa (in which the dominant circulating viral strain was wild-type); however, the clinical arm (PROTECT-Patient) occurred during the second wave of COVID-19 in South Africa (in which the dominant viral strain was the Beta variant). The study was ended prematurely when it was discovered by colleagues at the NICD that CCP collected from donors infected with the "wild-" type virus had poor neutralising capacity against the widely circulating Beta variant [29]. SANBS and WCBS operate independently of the hospitals, national laboratory services and research-based COVID-19 surveillance programmes and therefore, definitive results of which COVID-19 variant our CCP donors were infected with was not available to us, thus hampering the success of this study.

More than half of the donors made only three or fewer donations. Although this may appear to be disappointing, neutralising antibody titres wane over time, and a robust CCP programme relies on regular donations from donors with high titre antibody levels. Testing of antibody titres at each donation allowed for donors with high titres to be actively recruited, while those with low titres were encouraged to donate whole blood or source plasma instead. This had dual benefits as the number of clinically acceptable, high titre CCP donations increased, and the routine donor pool, which was extremely constrained due to lockdown effects, also increased.

5. Conclusions

Despite an intense international effort, there is still a lack of both definitive treatment options and broad access to preventative options for COVID-19. The therapeutic role of CCP is yet to be fully determined, with studies showing futility [10,11,13,30] and benefit [14,16,31] in different clinical settings. The CCP dose, timing, titre, donor and patient population are still of debate, and the emergence of variants of concern complicates the testing and selection of CCP. However, what is proven is the ability of blood services to rapidly implement a large-scale nationwide CCP collection, testing, processing and release programme. Blood services are uniquely positioned to implement CCP programmes by leveraging already established systems. The skills acquired and systems implemented to achieve this programme are likely to benefit SANBS and WCBS in the current and future pandemics. In addition, this programme has created a lasting legacy of new working partnerships, which have opened new avenues of collaboration for future projects.

Author Contributions: K.v.d.B. and M.V. designed and directed the study and provided overall supervision of the study; T.N.G., C.H., C.N., A.S., J.C., C.M., M.K., J.J., I.P. implemented the study; T.M.-G. and P.L.M. performed testing and analysis of SARS-Cov-2 antibody levels, P.L.M. and T.M.-G. provided analysis and interpretation of SARS-Cov-2 testing results; R.S. provided data and statistical analysis; T.B. and J.K. provided IT implementation; U.J. and D.N. performed, analysed and interpreted anti-HLA and anti-HNA antibody results; all authors provided critical feedback and helped shape the research, analysis and manuscript. All authors have read and agreed to the published version of the manuscript.

Funding: The research reported in this publication was supported by The ELMA South Africa Foundation (20-ESA011), Allan & Gill Gray Philanthropy LIMITED and the South African Medical Research Council with funds received from the Department of Science and Innovation. K.v.d.B. is supported in part by the NIH Fogarty International Center training grant 1D43-TW010345. P.L.M. is supported by the South African Research Chairs Initiative of the Department of Science and Innovation and National Research Foundation of South Africa (Grant No 98341). P.L.M. and T.M.-G. are supported by the SA Medical Research Council SHIP program.

Institutional Review Board Statement: The study was conducted according to the guidelines of the Declaration of Helsinki, and approved by the Human Research Ethics Committee of the South African National Blood Service (protocol code 2019/0519 on the 21 April 2020).

Informed Consent Statement: Informed consent was obtained from all subjects involved in the study.

Data Availability Statement: The data presented in this study are available on request from the corresponding author. The data are not publicly available due to SANBS data privacy policy.

Acknowledgments: SARS-CoV-2 Spike ELISA and neutralising antibody testing was performed by Zanele Makhado, Frances Ayres, Tandile Hermanus and Prudence Kgagudi.

Conflicts of Interest: The authors declare no conflict of interest.

References

1. Casadevall, A.; Pirofski, L.A. The convalescent sera option for containing COVID-19. *J. Clin. Investig.* **2020**, *1304*, 1545–1548. [CrossRef]
2. Mair-Jenkins, J.; Saavedra-Campos, M.; Baillie, J.K. The effectiveness of convalescent plasma and hyperimmune immunoglobulin for the treatment of severe acute respiratory infections of viral etiology: A systematic review and exploratory meta-analysis. *J. Infect. Dis.* **2015**, *2111*, 80–90. [CrossRef]
3. Sahr, F.; Ansumana, R.; Massaquoi, T.; Idriss, B.; Sesay, F.; Lamin, J.; Baker, S.; Nicol, S.; Conton, B.; Johnson, W.; et al. Evaluation of convalescent whole blood for treating Ebola Virus Disease in Freetown, Sierra Leone. *J. Infect.* **2017**, *743*, 302–309. [CrossRef]
4. Cheng, Y.; Wong, R.; Soo, Y.; Wong, W.; Lee, C.; Ng, M.; Chan, P.; Wong, K.; Leung, C.; Cheng, G. Use of convalescent plasma therapy in SARS patients in Hong Kong. *Eur. J. Clin. Microbiol. Infect. Dis.* **2005**, *241*, 44–46. [CrossRef] [PubMed]
5. Casadevall, A.; Dadachova, E.; Pirofski, L.A. Passive antibody therapy for infectious diseases. *Nat. Rev. Microbiol.* **2004**, *29*, 695–703. [CrossRef] [PubMed]
6. Piechotta, V.; Iannizzi, C.; Chai, K.L.; Valk, S.J.; Kimber, C.; Dorando, E.; Monsef, I.; Wood, E.M.; Lamikanra, A.A.; Roberts, D.J.; et al. Convalescent plasma or hyperimmune immunoglobulin for people with COVID-19: A living systematic review. *Cochrane Database Syst. Rev.* **2021**, *5*, CD013600. [PubMed]
7. Duan, K.; Liu, B.; Li, C. Effectiveness of convalescent plasma therapy in severe COVID-19 patients. *Proc. Natl. Acad. Sci. USA* **2020**, *11717*, 9490–9496. [CrossRef] [PubMed]
8. Chen, L.; Xiong, J.; Bao, L.; Shi, Y. Convalescent plasma as a potential therapy for COVID-19. *Lancet Infect. Dis.* **2020**, *204*, 398–400. [CrossRef]
9. Shen, C.; Wang, Z.; Zhao, F.; Yang, Y.; Li, J.; Yuan, J.; Wang, F.; Li, D.; Yang, M.; Xing, L.; et al. Treatment of 5 Critically Ill Patients With COVID-19 With Convalescent Plasma. *JAMA* **2020**, *32316*, 1582–1589. [CrossRef] [PubMed]
10. Horby, P.W.; Estcourt, L.; Peto, L.; Emberson, J.R.; Staplin, N.; Spata, E.; Pessoa-Amorim, G.; Campbell, M.; Roddick, A.; Brunskill, N.E.; et al. On behalf of the RECOVERY Collaborative Group. Convalescent plasma in patients admitted to hospital with COVID-19: A randomised controlled, open-label, platform trial. *Lancet* **2021**, *39710289*, 2049–2059.
11. Agarwal, A.; Mukherjee, A.; Kumar, G.; Chatterjee, P.; Bhatnagar, T.; Malhotra, P. Convalescent plasma in the management of moderate COVID-19 in adults in India: Open label phase II multicentre randomised controlled trial (PLACID Trial). *BMJ* **2020**, *371*, m3939. [CrossRef]
12. O'Donnell, M.R.; Grinsztejn, B.; Cummings, M.J.; Justman, J.E.; Lamb, M.R.; Eckhardt, C.M.; Philip, N.M.; Cheung, Y.K.; Gupta, V.; João, E.; et al. A randomized double-blind controlled trial of convalescent plasma in adults with severe COVID-19. *J. Clin. Investig.* **2021**, *131*, e150646. [CrossRef]
13. Li, L.; Zhang, W.; Hu, Y.; Tong, X.; Zheng, S.; Yang, J.; Kong, Y.; Ren, L.; Wei, Q.; Mei, H.; et al. Effect of Convalescent Plasma Therapy on Time to Clinical Improvement in Patients With Severe and Life-threatening COVID-19: A Randomized Clinical Trial. *JAMA* **2020**, *3245*, 460–470. [CrossRef]
14. Libster, R.; Pérez Marc, G.; Wappner, D.; Coviello, S.; Bianchi, A.; Braem, V.; Esteban, I.; Caballero, M.T.; Wood, C.; Berrueta, M.; et al. Early High-Titer Plasma Therapy to Prevent Severe COVID-19 in Older Adults. *N. Engl. J. Med.* **2021**, *3847*, 610–618. [CrossRef]
15. Gonzalez, S.E.; Regairaz, L.; Salazar, M.; Ferrando, N.; Gonzalez, V.; Ramos, P.C.; Pesci, S.; Vidal, J.M.; Kreplak, N.; Estenssoro, E. Timing of convalescent plasma administration and 28-day mortality for COVID-19 pneumonia. *medRxiv* **2021**. [CrossRef]
16. Hueso, T.; Pouderoux, C.; Péré, H.; Beaumont, A.-L.; Raillon, L.-A.; Ader, F.; Chatenoud, L.; Eshagh, D.; Szwebel, T.-A.; Martinot, M.; et al. Convalescent plasma therapy for B-cell-depleted patients with protracted COVID-19. *Blood* **2020**, *13620*, 2290–2295. [CrossRef] [PubMed]
17. Bloch, E.M.; Shoham, S.; Casadevall, A.; Sachais, B.S.; Shaz, B.; Winters, J.L.; Van Buskirk, C.; Grossman, B.J.; Joyner, M.; Henderson, J.P.; et al. Deployment of convalescent plasma for the prevention and treatment of COVID-19. *J. Clin. Investig.* **2020**, *1306*, 2757–2765. [CrossRef] [PubMed]
18. Bloch, E.M.; Goel, R.; Wendel, S.; Burnouf, T.; Al-Riyami, A.Z.; Ang, A.L.; DeAngelis, V.; Dumont, L.J.; Land, K.; Lee, C.K.; et al. Guidance for the procurement of COVID-19 convalescent plasma: Differences between high- and low-middle-income countries. *Vox Sang.* **2021**, *1161*, 18–35. [CrossRef] [PubMed]
19. South African Government COVID-19 about Alert System. Available online: https://www.gov.za/covid-19/about/about-alert-system (accessed on 26 July 2021).
20. U.S. Department of Health and Human Services Food and Drug Administration. Investigational COVID-19 Convalescent Plasma Guidance for Industry. Updated 11th February 2021. Available online: https://www.fda.gov/media/136798/download. (accessed on 6 July 2021).
21. Guidelines for Quarantine and Isolation in Relation to COVID-19 Exposure and Infection. Available online: https://www.nicd.ac.za/wp-content/uploads/2020/05/Guidelines-for-Quarantine-and-Isolation-in-relation-to-COVID-19.pdf (accessed on 26 July 2021).

22. Amanat, F.; Stadlbauer, D.; Strohmeier, S.; Nguyen, T.H.; Chromikova, V.; McMahon, M.; Jiang, K.; Arunkumar, G.A.; Jurczyszak, D.; Polanco, J.; et al. A serological assay to detect SARS-CoV-2 seroconversion in humans. *Nat. Med.* **2020**, *267*, 1033–1036. [CrossRef] [PubMed]
23. Moyo-Gwete, T.; Madzivhandila, M.; Makhado, Z.; Ayres, F.; Mhlanga, D.; Oosthuysen, B.; Lambson, B.E.; Kgagudi, P.; Tegally, H.; Iranzadeh, A.; et al. Cross-Reactive Neutralizing Antibody Responses Elicited by SARS-CoV-2 501Y.V2 (B.1.351). *N. Engl. J. Med.* **2021**, *38422*, 2161–2163. [CrossRef]
24. Legros, V.; Denolly, S.; Vogrig, M.; Boson, B.; Siret, E.; Rigaill, J.; Pillet, S.; Grattard, F.; Gonzalo, S.; Verhoeven, P.; et al. A longitudinal study of SARS-CoV-2-infected patients reveals a high correlation between neutralizing antibodies and COVID-19 severity. *Cell. Mol. Immunol.* **2021**, *182*, 318–327. [CrossRef] [PubMed]
25. Valdivia, A.; Torres, I.; Latorre, V.; Francés-Gómez, C.; Albert, E.; Gozalbo-Rovira, R.; Alcaraz, M.J.; Buesa, J.; Rodríguez-Díaz, J.; Geller, R.; et al. Inference of SARS-CoV-2 spike-binding neutralizing antibody titers in sera from hospitalized COVID-19 patients by using commercial enzyme and chemiluminescent immunoassays. *Eur. J. Clin. Microbiol. Infect. Dis.* **2021**, *403*, 485–494. [CrossRef] [PubMed]
26. Simtong, P.; Sudwilai, Y.; Cheunta, S.; Leelayuwat, C.; Romphruk, A.V. Prevalence of leucocyte antibodies in non-transfused male and female platelet apheresis donors. *Transfus. Med.* **2021**, *313*, 186–192. [CrossRef]
27. Triulzi, D.J.; Kleinman, S.; Kakaiya, R.M.; Busch, M.P.; Norris, P.J.; Steele, W.R.; Glynn, S.A.; Hillyer, C.D.; Carey, P.; Gottschall, J.L.; et al. The effect of previous pregnancy and transfusion on HLA alloimmunization in blood donors: Implications for a transfusion-related acute lung injury risk reduction strategy. *Transfusion* **2009**, *499*, 1825–1835. [CrossRef]
28. Regan, L.; Braude, P.R.; Hill, D.P. A prospective study of the incidence, time of appearance and significance of anti-paternal lymphocytotoxic antibodies in human pregnancy. *Hum. Reprod.* **1991**, *62*, 294–298. [CrossRef]
29. Wibmer, C.K.; Ayres, F.; Hermanus, T.; Madzivhandila, M.; Kgagudi, P.; Oosthuysen, B.; Lambson, B.E.; De Oliveira, T.; Vermeulen, M.; Van der Berg, K.; et al. SARS-CoV-2 501Y.V2 escapes neutralization by South African COVID-19 donor plasma. *Nat. Med.* **2021**, *274*, 622–625. [CrossRef]
30. Simonovich, V.A.; Burgos Pratx, L.D.; Scibona, P.; Beruto, M.V.; Vallone, M.G.; Vázquez, C.; Savoy, N.; Giunta, D.H.; Pérez, L.G.; Sánchez, M.d.L.; et al. A Randomized Trial of Convalescent Plasma in COVID-19 Severe Pneumonia. *N. Engl. J. Med.* **2021**, *3847*, 619–629. [CrossRef] [PubMed]
31. Tworek, A.; Jaroń, K.; Uszyńska-Kałuża, B.; Rydzewski, A.; Gil, R.; Deptała, A.; Franek, E.; Wójtowicz, R.; Życińska, K.; Walecka, I.; et al. Convalescent plasma treatment is associated with lower mortality and better outcomes in high-risk COVID-19 patients-propensity-score matched case-control study. *Int. J. Infect. Dis.* **2021**, *105*, 209–215. [CrossRef] [PubMed]

 viruses

MDPI

Article

In Vivo Study of Aerosol, Droplets and Splatter Reduction in Dentistry

Naeemah Noordien [1], Suné Mulder-van Staden [2,*] and Riaan Mulder [3]

[1] Paediatric Dentistry, The University of the Western Cape, Cape Town 7530, South Africa; nnoordien@uwc.ac.za
[2] Oral Medicine, Periodontology and Implantology Department, The University of the Western Cape, Cape Town 7530, South Africa
[3] Restorative Dentistry, The University of the Western Cape, Cape Town 7530, South Africa; rmulder@uwc.ac.za
* Correspondence: smuldervanstaden@uwc.ac.za; Tel.: +27-21-937-3000

Abstract: Oral health care workers (OHCW) are exposed to pathogenic microorganisms during dental aerosol-generating procedures. Technologies aimed at the reduction of aerosol, droplets and splatter are essential. This in vivo study assessed aerosol, droplet and splatter contamination in a simulated clinical scenario. The coolant of the high-speed air turbine was colored with red concentrate. The red aerosol, droplets and splatter contamination on the wrists of the OHCW and chests of the OHCW/volunteer protective gowns, were assessed and quantified in cm^2. The efficacy of various evacuation strategies was assessed: low-volume saliva ejector (LV) alone, high-volume evacuator (HV) plus LV and an extra-oral dental aerosol suction device (DASD) plus LV. The Kruskal–Wallis rank-sum test for multiple independent samples with a post-hoc test was used. No significant difference between the LV alone compared to the HV plus LV was demonstrated ($p = 0.372059$). The DASD combined with LV resulted in a 62% reduction of contamination of the OHCW. The HV plus LV reduced contamination by 53% compared to LV alone ($p = 0.019945$). The DASD demonstrated a 50% reduction in the contamination of the OHCWs wrists and a 30% reduction in chest contamination compared to HV plus LV. The DASD in conjunction with LV was more effective in reducing aerosol, droplets and splatter than HV plus LV.

Keywords: SARS-CoV-2; dental aerosol-generating procedures; extra-oral suction; high-volume evacuation; low-volume saliva ejector; splatter; aerosol

Citation: Noordien, N.; Mulder-van Staden, S.; Mulder, R. In Vivo Study of Aerosol, Droplets and Splatter Reduction in Dentistry. *Viruses* **2021**, *13*, 1928. https://doi.org/10.3390/v13101928

Academic Editors: Burtram C. Fielding and Georgia Schäfer

Received: 9 September 2021
Accepted: 22 September 2021
Published: 25 September 2021

Publisher's Note: MDPI stays neutral with regard to jurisdictional claims in published maps and institutional affiliations.

1. Introduction

The dental environment is unique in the high risk it poses for the transmission of infectious agents [1,2]. Oral health care workers (OHCW) can potentially be exposed to numerous pathogens (such as viruses, bacteria and fungi) that are present in the oral cavity and respiratory tract of patients [3]. OHCWs are amongst the highest risk group for disease contamination by aerosols, droplets and splatter [4]. The origin of airborne contaminants in a dental setting could be in the form of saliva, dental instruments, patient respiratory sources and the oral cavity [5]. Water combined with compressed air produces aerosol, droplets and splatter, which become contaminated by the oral cavity [6]. Aerosol-generating procedures (AGPs) produce a mixture of aerosol, droplets and splatter containing blood and saliva with various microorganisms [7]. This creates a working environment with a high potential of disease transmission [8]. The literature has demonstrated numerous sources of aerosol, droplet and splatter production in the dental environment—such as ultrasonic scalers and high-speed air turbines [9,10]

Dental lasers are also considered aerosol-producing devices due to the generation of a plume during procedures [11]. During laser procedures, a high-efficiency particulate filtration respirator (N99/FFP3 respirator) with a filter efficiency of 99.75% at 0.1 µm has been recommended [11]. The literature has demonstrated that the aerosols, droplets and splatter produced in large quantities during daily dental procedures are contaminated with pathogens [12]. These contaminated aerosols and droplets can remain suspended for extended periods of time before entering the respiratory tract or settling on surfaces [13]. Aerosols generated during dental procedures may remain suspended in the air for several hours and can spread up to 3 m from the source [9]. Aerosolized particles generated by dental equipment range from aerosol to droplets to splatter (0.001 to 50 μm). Particle sizes influence the time of suspension and settling of these aerosolized particles. Particles greater than 100 μm can be classified as splatter and settle quickly on surrounding surfaces and the floor. Droplet particle sizes that are smaller than 50 μm remain suspended in the air for extended periods [14].

Studies have demonstrated that aerosolized particles of 1 μm consist of sufficient volume to harbor a variety of respiratory pathogens and allow disease transmission. This has been demonstrated when pathogen transmission with the measles virus (50–500 nm) [15], influenza virus (100 nm to 1 μm) [16] and Mycobacterium tuberculosis (1–3 μm) has been studied. The current SARS-CoV-2 coronavirus pandemic brought to the forefront the concern of infection spread and transmission in the dental setting [17]. The pandemic resulted in a nearly complete halt in dental treatment across the world due to various lockdown regulations. In Italy and many other countries, dental treatment was limited to urgency and emergency treatment that could not be postponed. Dentistry performed during the initial stages of the SARS-CoV-2 pandemic was highlighted by an Italian study. It was reported that 69.5% of dentists managed dental emergencies recognized by the American Dental Association. Further, 68.2% of the dentists reported a fear of contracting SARS-CoV-2 after treatments were performed [18]. The importance of reducing the exposure of OHCW and the dental environment to aerosols, droplets and splatter generated during dental procedures has now become of even greater concern. Routes of transmission of SARS-CoV-2 coronavirus in humans include contact transmission (by means of contact with oral, nasal and eye mucosal membranes) and direct transmission (by means of coughing, sneezing and droplet inhalation transmission) [13]. Patients that are SARS-CoV-2 coronavirus positive, however asymptomatic, will inevitably present to dental practices for treatment. Studies have demonstrated that these asymptomatic SARS-CoV-2 coronavirus patients, as well as those recovering from acute illness, continue to shed significant amounts of the virus [19]. Studies have also demonstrated that symptomatic and asymptomatic SARS-CoV-2 coronavirus-positive patients presented with very similar viral loads [20]. Thus, the potential for generated droplets and aerosolized particles from these patients during procedures are at high risk for contaminating the air and surfaces of the entire dental practice [13]. The aim should thus be to reduce the amount of aerosol, droplets and splatter to an absolute minimum during this SARS-Cov-2 coronavirus pandemic [5]. The aim of this study was to assess the efficacy of a novel dental aerosol suction device (DASD) in reducing aerosol, droplets and splatter contamination.

2. Materials and Methods

2.1. Experimental Setup

The experiment was carried out in a 16 m^2 dental surgery. A simulated clinical scenario was created with the high-speed air turbine. The OHCW (SMVS) positioned the air turbine above the right mandibular molar (tooth 46, FDI World Dental Federation notation) of a live volunteer (RM). The simulated clinical scenario was defined as the dental bur in the high-speed air turbine, directed 1 mm away from the central fissure of tooth 46 of the volunteer (RM) for 5 min. This would simulate the time spent on a cavity preparation, based on authors reporting full crown preparations in six minutes [21]. The tooth of the volunteer never made contact with the diamond bur during the simulation. The assistant and independent researcher (NN) held the Dental Aerosol Suction Device (DASD), high-volume evacuation and low-volume saliva ejector as per the study design. This simulated clinical scenario facilitated the in vivo assessment of the novel Dental Aerosol Suction Device (DASD). The OHCW was positioned 40 cm from the volunteer oral cavity holding the high-speed air turbine in position. The assistant was 40 cm from the volunteer oral cavity holding the low-volume saliva ejector (LV) and other devices that were assessed in position.

2.2. Equipment Used

The DASD device is patented in the United Kingdom under registration 6119833. The international design classification cover Class 24: Medical and laboratory equipment with subclass 02: Medical instruments, instruments and tools for laboratory use. The manufacturer ensures the CE marking. The DASD device is manufactured from a durable Nylon material under the ISO 13485. The DASD is autoclavable and capable of attaching to the high-volume evacuation adapter of the dental unit. The DASD device does not require additional motors or power sources to operate. The DASD design optimizes the catchment area, ergonomics of holding the device and aerodynamics to optimize the suction volume when larger than 300 L/min. The equipment utilized in this study included: DASD: Dental aerosol suction device (The University of the Western Cape, Cape Town, South Africa) (Figure 1); low-volume saliva ejector (LV) (Removable 6IN clear 22810148, Henry Schein, Johannesburg, South Africa); 11 mm high-volume straight evacuation tip (HV) (Saliva Ejec white 11 mm 078110, Henry Schein, Johannesburg, South Africa); Aspijet7 mobile suction unit at 400 L/min suction when the low-volume and high-volume suction adapters are open in full (Cattani ESAM, Worcestershire, UK) Durr suction vacuum and airflow rate volume gauge (Durr Dental SE, Bietigheim-Bissingen, Germany); high-speed air turbine (Pana-Max PLUS, NSK, Kanuma, Japan) (340,000 rpm), fitted with an inverted cone diamond bur (FG320R-5, Kerr, Brea, CA, USA) directed at the central fissure of the tooth 46 molar. Coolant flow rates were adjusted to 15 mL/min for the high-speed air turbine.

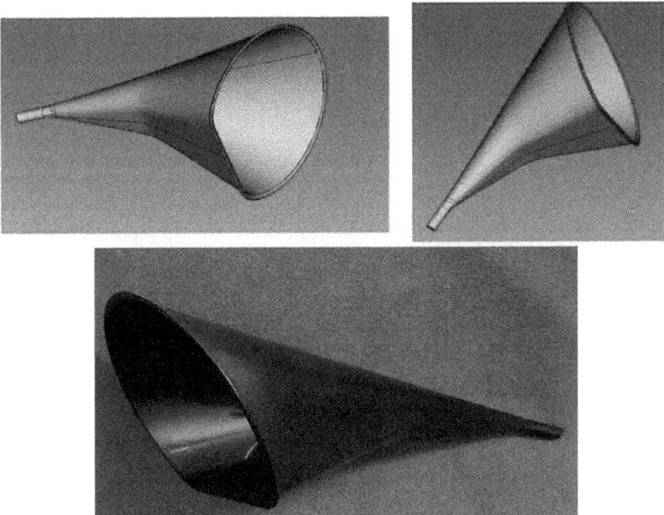

Figure 1. Dental Aerosol Suction Device (DASD).

2.3. Measurement Method

The long-sleeved fluid-resistant protective gowns were assessed under 1.75 × magnification (Start International, Dallas, TX, USA) for red contamination from the aerosol, droplets and splatter that occurred during the simulated clinical scenario. This clinical scenario was replicated in triplicate, each time with new long-sleeved fluid-resistant protective gowns. The areas contaminated from aerosol, droplets and splatter presented as red contaminated areas on the fluid-resistant protective gowns and were quantified under magnification in cm^2 from an overlaid A4 paper size clear transparency with 1 cm^2 blocks. The cm^2 blocks were counted until a 1 cm^2 block with no visible contamination was encountered—demarcating the end of contamination. The red color of the coolant was derived from red concentrate added in a concentration of 50 mL to 1000 mL distilled water coolant. The water line was purged with this prepared coolant prior to starting the in vivo study. The surfaces evaluated for contamination from aerosol, droplets and splatter were the chest of the volunteer (RM), the wrists and the chest of the OHCW (SMVS) (Figure 2). The chest of the OHCW was evaluated from a position 20 cm below the collar, as this is where the protective shield ends. The assessment area for the wrists was demarcated as the circumferential area of 20 cm (of the protective gown sleeves). The chest of the volunteer was demarcated as 40 cm starting at the collar. The width of the chest area was 30 cm. The test groups consisted of the control group (LV): low-volume saliva ejector on the low-volume evacuation adapter alone in the left lingual fossa; the high-volume evacuation (HV) consisted of the conventional high-volume evacuation 11 mm diameter tip 1 cm away from the high-speed air turbine head attached to the high-volume adapter and the LV as described; followed by the Dental Aerosol Suction Device (DASD) attached to the high-volume suction adapter with the LV as described. The DASD device is held by the assistant 10 cm away from the corner of the mouth in the 5 O'clock position. The air conditioner system in the dental surgery was off, and no natural ventilation was present with doors and windows closed.

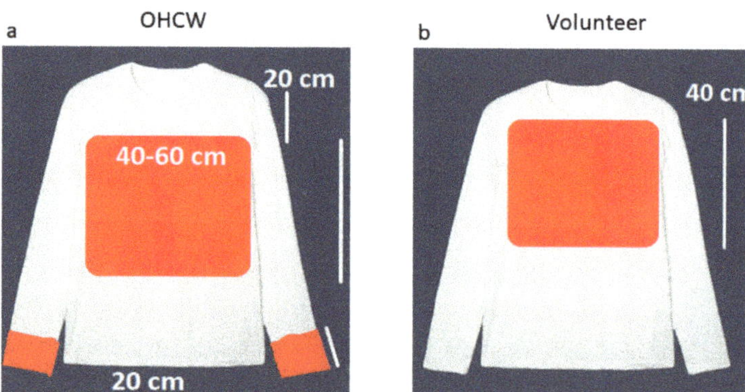

Figure 2. Long-sleeved fluid-resistant protective gown surfaces assessed for contamination (**a**) OHCW chest and wrists and (**b**) volunteer chest.

2.4. Statistical Analysis

The data were analyzed using R Core Team (2013); (R: A language and environment for statistical computing. R Foundation for Statistical Computing, Vienna, Austria). The mean and standard deviation was calculated for the aerosol, droplets and splatter produced during the simulated clinical scenarios with the various aerosol reducing equipment. The Kruskal–Wallis rank-sum test for multiple independent samples was used. The Tukey–Kramer (Nemenyi) post-hoc pairwise multiple comparison tests were conducted to assess the significant differences. A p-value ≤ 0.05 was considered statistically significant.

3. Results

The Kruskal–Wallis rank-sum test for multiple independent samples resulted in a p-value = 0.027324. The post-hoc pairwise multiple comparison test was conducted to discern which of the pairs have significant differences. The input data reveals no ties in the ranks; therefore, p-value adjustments were not applicable. Tukey–Kramer (Nemenyi) p-values indicated a significant difference between the contamination of the volunteer chest, oral health care workers' chest and wrists when LV alone was used compared to the

DASD plus LV (Table 1). There was no significant difference between the LV alone and HV plus LV ($p = 0.372059$).

Table 1. Collection of aerosol and splatter on different areas of the OHCW and volunteer.

Device Assessed	Volunteer Chest (cm^2)	OHCW Chest below Shield (cm^2)	OHCW Wrists (cm^2)
LV only	105 (±6.02)	357 (±4)	118 (±4)
HV + LV	55 (±5)	192 (±4.58)	71 (±2.64)
DASD + LV	25 (±2) *	133 (±2) *	35 (±1) *

*, $p = 0.019945$.

4. Discussion

The oral cavity harbors multiple pathogens with the potential of infecting OHCW during aerosol-generating dental procedures [22]. An analysis performed by the Alberta Federation of Labor established that OHCW is listed amongst the top 100 occupations with the highest risk of SARS-CoV-2 coronavirus exposure. This analysis also stratified exposure risk as follows: dental technologists 62.5% risk, dental assistants 97.5% risk, dentists 97% risk and dental hygienists and therapists carry 100% risk [23]. The literature currently does not include an assortment of in vivo data on the effectiveness of current devices to reduce aerosol, droplet and splatter contamination in dentistry [5]. Low-volume saliva ejectors (LV) alone have been deemed insufficient for aerosol-generating procedures in dental practice. Studies evaluating LV alone in the simulated clinical scenario detected very high concentrations of ultrafine aerosol particles (<10 μm) [21].

Currently, the use of high-volume evacuation and rubber dams are techniques aimed at minimizing microbial loaded dental aerosol and droplet contamination [12]. Studies have concluded that the use of rubber dams does reduce the microbial contamination of the operator and surrounding dental environment [24]. However, rubber dams have also been demonstrated to be associated with increased contamination of sterile head scarfs, as the rubber dam increases the average particle size [25]. Han et al. (2021) concluded that more studies are required in order to test the efficacy of aerosol, droplet and splatter reduction with dental high-volume evacuation devices [7]. Some authors have suggested that extra-oral motor-driven suction devices are good tools to reduce aerosol during dental treatment [8,26,27]. Study data regarding a reduction in aerosol particle movement demonstrated it in a dental setting when using a motor-driven extra-oral suction device [17]. The Isolite® device is an example of an extra-oral suction device. A study evaluating the Isolite® device assessed variables such as plaque, saliva, patient and operator position. This study demonstrated no significant difference in aerosol, droplet and splatter with regards to colony-forming units with the Isolite® device, compared to low-volume saliva ejector alone [28].

Air purifiers have also been utilized in an attempt to reduce aerosol spread. The use of air purifiers was demonstrated to be insignificant in clinical settings where the cubicles are open, and multiple dental chairs are positioned in close proximity [17]. Natural air ventilation has been shown to assist in the dissemination of particulates away from areas where it would have settled. A study demonstrated that 1 m from the OHCW and 0.5 m from the saliva evacuation unit, a greater volume of particulate settled when low-volume saliva ejector was used with natural ventilation, compared to low-volume saliva ejector alone [29]. This study evaluated the aerosol, droplet and splatter reduction that could be achieved with a relatively inexpensive device (DASD). This device is not motor driven, does not require an additional power source and can be directly connected to the dental chair high-volume evacuation adapter. A study discussing the design of aerosol suction devices concluded that an optimal device size and shape were required to ensure aerosol reduction [5]. The DASD device is unique in its design and large catchment area, which optimizes the available high-volume suction capacity of the dental unit (above 300 L/min).

Larger, more expensive extra-oral high-volume suction devices have integrated suction motors and air filters. A recent study evaluated an extra-oral high-volume device positioned superior-perpendicular to the mannequin's oral cavity. This extra-oral device achieved a significant reduction in operator contamination compared to high-volume evacuation [30]. A study identified the drilling side and corresponding location to be the ideal position for high-volume evacuation tips [21]. An advantage of the DASD design and large catchment area is that the device positioning for optimal aerosol and splatter reduction is achievable from multiple positions. The DASD device can be positioned on the chest of the patient anterior to the oral cavity (termed anterior-perpendicular in the 6 O'clock position) and laterally (5 and 7 O'clock positions) (Figure 3). This allows versatile positioning of the DASD device, which does not interfere with the working field of the operator and the position of the overhead light. The DASD is recommended to be positioned 10 to 15 cm from the working area. This provides the advantage of not impeding the OHCW's field of vision, compared to high-volume suction that needs to be positioned up to 1 cm adjacent to the working field.

Figure 3. Positioning possibilities of the Dental Aerosol Suction Device.

A recent study assessed and compared commercially available high-volume evacuation and an extra-oral motor-driven vacuum system. The authors concluded that the low-volume saliva ejector alone was not adequate during aerosol-generating procedures. The authors also demonstrated no significant difference regarding aerosol contamination of OHCW when low-volume saliva ejector or high-volume evacuation was combined with a motor-driven extra-oral vacuum system [5]. The current DASD study demonstrated a 53% reduction in aerosol, droplet and splatter contamination of an OHCW when a high-volume evacuation was used in conjunction with a low-volume saliva ejector, compared to a low-volume saliva ejector alone. The DASD demonstrated a 62% reduction in aerosol, droplet and splatter contamination of an OHCW when used in conjunction with a low-volume saliva ejector. The aerosol, droplet and splatter contamination of the chest of the patient and the OHCW is a concerning factor. Studies have demonstrated that aerosols, droplets and splatters have been identified as far as 1.2 m from the source [7]. A study demonstrated that high-volume evacuation markedly reduced aerosol and droplet at the patient's position of the oral cavity and at the level of the clinician when compared to low-volume saliva ejectors [5]. Contamination of the chest and wrist of OHCW during aerosol-generating procedures carries a high risk of cross-contamination and self-inoculation. The DASD achieved a 50% reduction in the contamination of the OHCW wrists and demonstrated a

30% reduction in the contamination of the OHCWs chest (DASD plus LV), compared to high-volume evacuation plus LV.

A study evaluating high-volume evacuation with high-volume evacuation plus extra oral motorized vacuum system achieved a 50% reduction in wrist contamination [30]. The DASD device plus LV achieved the same results utilizing the chair suction alone (at ≥300 L/min), without the need for additional motorized devices and power sources. Studies have demonstrated the presence of SARS-CoV-2 in the aerosol produced during patient exhalation [31,32]. During aerosol generation, the high-volume evacuation tip only operates for limited time frames during close approximation to the tooth during procedures, which reduces the efficacy of minimizing aerosols exhaled by the patient [5]. The DASD has the advantage of a large catchment area and can be placed statically to continuously eliminate aerosols produced during patient exhalation from the 6 O'clock position.

5. Conclusions

SARS-CoV-2 coronavirus has brought the importance of infection control and aerosol reduction in dentistry to the forefront. However, the importance of creating a safe working environment in dentistry should be a priority at all times. All OHCW should be practicing in a manner that ensures infection control (with a focus on the control of aerosol, droplets and splatter) and strictly adhere to disinfection protocols.

The DASD device presents as a cost-effective option to reduce aerosol, droplets and splatter, compared to current more expensive extra-oral evacuation systems. Greater levels of aerosol, droplets and splatter reduction to the wrists of the OHCW, as well as the chest of the OHCW and volunteer, was achieved with this device. The DASD enables aerosol, droplets and splatter reduction from multiple positions without impeding OHCW visibility and accessibility to the oral cavity.

6. Limitations

The contamination of the assistant was not recorded in this study. The dental bur positioned 1 cm away from the tooth fissure might not cover all clinical scenarios of cavity preparation, as the occlusal aspect of the tooth is essentially a flat, smooth surface. In a cavity preparation, the coolant would be directed into and then out of the cavity, possibly at a different trajectory compared to the simulated clinical scenario. Contamination was quantified in cm^2 blocks, irrespective of the degree of red contamination per cm^2. The addition of the red concentrate to the distilled water (similar to studies utilizing sodium fluorescein) could potentially alter the water tension, flow of the red coolant and aerosol, droplet and splatter sizes. The change in water tension with the red concentrate was not compared to distilled water prior to the study. A recommendation for future studies includes the assessment of aerosol generation during procedures, such as tooth polishing, surgical interventions and ultrasonic scaling. Further research could be conducted with DASD and a low-volume saliva ejector to indicate the efficacy of the device in those procedures with an electronic particle sensor.

Author Contributions: Methodology: R.M. and S.M.-v.S. Conceptualization, investigation, writing original draft preparation and writing—review and editing manuscript: N.N. and S.M.-v.S., N.N. served as an independent observer and maintained the study parameters. All authors have read and agreed to the published version of the manuscript.

Funding: This research received external funding from the Author Incentive Fund of R.M. grant number 111388 and the Technology Innovation Agency, Pretoria, South Africa grant number UWC032-TSF. The University of the Western Cape, Cape Town, South Africa, is the owner of the DASD patents in the United Kingdom and South Africa.

Institutional Review Board Statement: The University of the Western Cape under BMREC research committee approved the research project: Dissemination of aerosols in dentistry during the use of an aerosol suction device; project number: BM20/5/2.

Informed Consent Statement: Informed consent for the volunteer was waived due to Riaan Mulder being the co-researcher and volunteer.

Data Availability Statement: Available upon request.

Conflicts of Interest: The author N.N. declares no conflict of interest. R.M. and S.M.-v.S. are the inventors of the DASD device.

References

1. Singla, D.; Singh, A.; Shiva Manjunath, R.; Bhattacharya, H.; Sarkar, A.; Chandra, N. Aerosol, a health hazard during ultrasonic scaling: A clinico-microbiological study. *Indian J. Dent. Res.* **2016**, *27*, 160–162. [CrossRef]
2. Szymańska, J. Dental bioaerosol as an occupational hazard in a dentist's workplace. *Ann. Agric. Environ. Med.* **2007**, *14*, 203–207.
3. Kampf, G.; Todt, D.; Pfaender, S.; Steinmann, E. Persistence of coronaviruses on inanimate surfaces and their inactivation with biocidal agents. *J. Hosp. Infect.* **2020**, *104*, 246–251. [CrossRef]
4. Laheij, A.; Kistler, J.; Belibasakis, G.; Välimaa, H.; de Soet, J. Healthcare-associated viral and bacterial infections in dentistry. *J. Oral. Microbiol.* **2012**, *4*, 17659. [CrossRef]
5. Matys, J.; Grzech-Leśniak, K. Dental Aerosol as a Hazard Risk for Dental Workers. *Materials* **2020**, *13*, 5109. [CrossRef]
6. Zemouri, C.; Volgenant, C.; Buijs, M.; Crielaard, W.; Rosema, N.; Brandt, B.; Laheij, A.; De Soet, J. Dental aerosols: Microbial composition and spatial distribution. *J. Oral. Microbiol.* **2020**, *12*, 1762040. [CrossRef]
7. Han, P.; Li, H.; Walsh, L.; Ivanovski, S. Splatters and Aerosols Contamination in Dental Aerosol Generating Procedures. *Appl. Sci.* **2020**, *11*, 1914. [CrossRef]
8. Senpuku, H.; Fukumoto, M.; Uchiyama, T.; Taguchi, C.; Suzuki, I.; Arikawa, K. Effects of Extraoral Suction on Droplets and Aerosols for Infection Control Practices. *Dent. J.* **2021**, *9*, 80. [CrossRef]
9. Veena, H.; Mahantesha, S.; Joseph, P.; Patil, S.; Patil, S. Dissemination of aerosol and splatter during ultrasonic scaling: A pilot study. *J. Infect. Public Health* **2015**, *8*, 260–265. [CrossRef]
10. Prospero, E.; Savini, S.; Annino, I. Microbial Aerosol Contamination of Dental Healthcare Workers' Faces and Other Surfaces in Dental Practice. *Infect. Control Hosp. Epidemiol.* **2003**, *24*, 139–141. [CrossRef]
11. Convissar, R. *Principles and Practice of Laser Dentistry*, 2nd ed.; Mosby Elsevier: St. Louis, MO, USA, 2015; pp. 1–328.
12. Wei, J.; Li, Y. Airborne spread of infectious agents in the indoor environment. *Am. J. Infect. Control* **2016**, *44*, S102–S108. [CrossRef]
13. Peng, X.; Xu, X.; Li, Y.; Cheng, L.; Zhou, X.; Ren, B. Transmission routes of 2019-nCoV and controls in dental practice. *Int. J. Oral Sci.* **2020**, *12*, 1–6. [CrossRef]
14. Singh, T.; Mabe, O. Occupational exposure to endotoxin from contaminated dental unit waterlines. *SADJ* **2009**, *64*, 10–12.
15. Liljeroos, L.; Huiskonen, J.; Ora, A.; Susi, P.; Butcher, S. Electron cryotomography of measles virus reveals how matrix protein coats the ribonucleocapsid within intact virions. *Proc. Natl. Acad. Sci. USA* **2011**, *108*, 18085–18090. [CrossRef]
16. Rossman, J.; Lamb, R. Influenza virus assembly and budding. *Virology* **2011**, *411*, 229–236. [CrossRef]
17. Makhsous, S.; Segovia, J.; He, J.; Chan, D.; Lee, L.; Novosselov, I.; Mamishev, A. Methodology for Addressing Infectious Aerosol Persistence in Real-Time Using Sensor Network. *Sensors* **2021**, *21*, 3928. [CrossRef]
18. Sinjari, B.; Rexhepi, I.; Santilli, M.; D'Addazio, G.; Chiacchiaretta, P.; Di Carlo, P.; Caputi, S. The Impact of COVID-19 Related Lockdown on Dental Practice in Central Italy—Outcomes of A Survey. *Int. J. Environ. Res. Public Health* **2020**, *17*, 5780. [CrossRef]
19. Chang, D.; Mo, G.; Yuan, X.; Tao, Y.; Peng, X.; Wang, F.; Xie, L.; Sharma, L.; Dela Cruz, C.; Qin, E. Time Kinetics of Viral Clearance and Resolution of Symptoms in Novel Coronavirus Infection. *Am. J. Respir. Crit. Care Med.* **2020**, *201*, 1150–1152. [CrossRef]
20. Zou, L.; Ruan, F.; Huang, M.; Liang, L.; Huang, H.; Hong, Z.; Yu, J.; Kang, M.; Song, Y.; Xia, J.; et al. SARS-CoV-2 Viral Load in Upper Respiratory Specimens of Infected Patients. *N. Eng. J. Med.* **2020**, *382*, 1177–1179. [CrossRef]
21. Balanta-Melo, J.; Gutiérrez, A.; Sinisterra, G.; Díaz-Posso, M.; Gallego, D.; Villavicencio, J.; Contreras, A. Rubber Dam Isolation and High-Volume Suction Reduce Ultrafine Dental Aerosol Particles: An Experiment in a Simulated Patient. *Appl. Sci.* **2020**, *10*, 6345. [CrossRef]
22. Zemouri, C.; De Soet, H.; Crielaard, W.; Laheij, A. A scoping review on bio-aerosols in healthcare and the dental environment. *PLoS ONE* **2017**, *12*, e0178007. [CrossRef]
23. Alberta Federation of Labour. Available online: https://www.afl.org/as_albertans_return_to_work_who_is_at_the_highest_risk_of_exposure_to_the_novel_coronavirus (accessed on 5 September 2021).
24. Cochran, M.; Miller, C.; Sheldrake, M. The efficacy of the rubber dam as a barrier to the spread of microorganisms during dental treatment. *JADA* **1989**, *119*, 141–144. [CrossRef]
25. Al-Amad, S.; Awad, M.; Edher, F.; Shahramian, K.; Omran, T. The effect of rubber dam on atmospheric bacterial aerosols during restorative dentistry. *J. Infect. Public Health* **2017**, *10*, 195–200. [CrossRef] [PubMed]
26. Shahdad, S.; Patel, T.; Hindocha, A.; Cagney, N.; Mueller, J.-D.; Sedoudi, N.; Morgan, C.; Din, A. The efficacy of an extraoral scavenging device on reduction of splatter contamination during dental aerosol generating procedures: An exploratory study. *Br. Dent. J.* **2020**, *11*, 1–10. [CrossRef]
27. Chavis, S.E.; Hines, S.E.; Dyalram, D.; Wilken, N.C.; Dalby, R.N. Can extraoral suction units minimize droplet spatter during a simulated dental procedure? *JADA* **2021**, *152*, 157–165. [CrossRef] [PubMed]

28. Holloman, J.; Mauriello, S.; Pimenta, L.; Arnold, R. Comparison of suction device with saliva ejector for aerosol and spatter reduction during ultrasonic scaling. *JADA* **2015**, *146*, 27–33. [CrossRef] [PubMed]
29. Rexhepi, I.; Mangifesta, R.; Santilli, M.; Guri, S.; Di Carlo, P.; D'Addazio, G.; Caputi, S.; Sinjari, B. Effects of Natural Ventilation and Saliva Standard Ejectors during the COVID-19 Pandemic: A Quantitative Analysis of Aerosol Produced during Dental Procedures. *Int. J. Environ. Res. Public Health* **2021**, *18*, 7472. [CrossRef]
30. Horsophonphong, S.; Chestsuttayangkul, Y.; Surarit, R.; Lertsooksawat, W. Efficacy of extraoral suction devices in aerosol and splatter reduction during ultrasonic scaling: A laboratory investigation. *J. Dent. Res. Dent. Clin. Dent. Prospects* **2021**, *15*, 197–202.
31. Leung, N.; Chu, D.; Shiu, E.; Chan, K.; McDevitt, J.; Hau, B.; Yen, H.; Li, Y.; Ip, D.; Peiris, J.; et al. Author Correction: Respiratory virus shedding in exhaled breath and efficacy of face masks. *Nat. Med.* **2020**, *26*, 981. [CrossRef]
32. Di Carlo, P.; Falasca, K.; Ucciferri, C.; Sinjari, B.; Aruffo, E.; Antonucci, I.; Di Serafino, A.; Pompilio, A.; Damiani, V.; Mandatori, D.; et al. Normal breathing releases SARS-CoV-2 into the air. *J. Med. Microbiol.* **2021**, *70*, 001328. [CrossRef]

Article

Patient and Clinical Factors at Admission Affect the Levels of Neutralizing Antibodies Six Months after Recovering from COVID-19

Xinjie Li [1,†], Ling Pang [2,†], Yue Yin [1], Yuqi Zhang [1], Shuyun Xu [3], Dong Xu [2,*] and Tao Shen [1,*]

1 Department of Microbiology and Infectious Disease Center, School of Basic Medical Sciences, Peking University, Beijing 100191, China; xinjieli@hsc.pku.edu.cn (X.L.); yuey@bjmu.edu.cn (Y.Y.); zhangyuqi_@bjmu.edu.cn (Y.Z.)

2 Department and Institute of Infectious Disease, Tongji Hospital, Tongji Medical College, Huazhong University of Science and Technology, Wuhan 430030, China; pangling000@163.com

3 Department of Respiratory and Critical Care Medicine, Key Laboratory of Pulmonary Diseases of Health Ministry, Tongji Hospital, Tongji Medical College, Huazhong University of Science and Technology, Wuhan 430030, China; sxu@hust.edu.cn

* Correspondence: xdong@tjh.tjmu.edu.cn (D.X.); taoshen@hsc.pku.edu.cn (T.S.); Tel.: +86-13-071-215-093 (D.X.); +86-10-828-05-070 (T.S.)

† These authors contributed equally to this work.

Abstract: The rate of decline in the levels of neutralizing antibodies (NAbs) greatly varies among patients who recover from Coronavirus disease 2019 (COVID-19). However, little is known about factors associated with this phenomenon. The objective of this study is to investigate early factors at admission that can influence long-term NAb levels in patients who recovered from COVID-19. A total of 306 individuals who recovered from COVID-19 at the Tongji Hospital, Wuhan, China, were included in this study. The patients were classified into two groups with high (NAb^{high}, $n = 153$) and low (NAb^{low}, $n = 153$) levels of NAb, respectively based on the median NAb levels six months after discharge. The majority (300/306, 98.0%) of the COVID-19 convalescents had detected NAbs. The median NAb concentration was 63.1 (34.7, 108.9) AU/mL. Compared with the NAb^{low} group, a larger proportion of the NAb^{high} group received corticosteroids (38.8% vs. 22.4%, $p = 0.002$) and IVIG therapy (26.5% vs. 16.3%, $p = 0.033$), and presented with diabetes comorbidity (25.2% vs. 12.2%, $p = 0.004$); high blood urea (median (IQR): 4.8 (3.7, 6.1) vs. 3.9 (3.5, 5.4) mmol/L; $p = 0.017$); CRP (31.6 (4.0, 93.7) vs. 16.3 (2.7, 51.4) mg/L; $p = 0.027$); PCT (0.08 (0.05, 0.17) vs. 0.05 (0.03, 0.09) ng/mL; $p = 0.001$); SF (838.5 (378.2, 1533.4) vs. 478.5 (222.0, 1133.4) µg/L; $p = 0.035$); and fibrinogen (5.1 (3.8, 6.4) vs. 4.5 (3.5, 5.7) g/L; $p = 0.014$) levels, but low SpO_2 levels (96.0 (92.0, 98.0) vs. 97.0 (94.0, 98.0)%; $p = 0.009$). The predictive model based on Gaussian mixture models, displayed an average accuracy of 0.7117 in one of the 8191 formulas, and ROC analysis showed an AUC value of 0.715 (0.657–0.772), and specificity and sensitivity were 72.5% and 67.3%, respectively. In conclusion, we found that several factors at admission can contribute to the high level of NAbs in patients after discharge, and constructed a predictive model for long-term NAb levels, which can provide guidance for clinical treatment and monitoring.

Keywords: coronavirus disease 2019 (COVID-19); severe acute respiratory syndrome coronavirus (SARS-CoV-2); neutralizing antibody (NAb); diabetes; corticosteroids

Citation: Li, X.; Pang, L.; Yin, Y.; Zhang, Y.; Xu, S.; Xu, D.; Shen, T. Patient and Clinical Factors at Admission Affect the Levels of Neutralizing Antibodies Six Months after Recovering from COVID-19. *Viruses* **2022**, *14*, 80. https://doi.org/10.3390/v14010080

Academic Editors: Burtram C. Fielding and Georgia Schäfer

Received: 12 November 2021
Accepted: 30 December 2021
Published: 2 January 2022

1. Introduction

The severe acute respiratory syndrome coronavirus type 2 (SARS-CoV-2) pandemic has caused havoc around the world. Immunity after recovery from Coronavirus disease 2019 (COVID-19), is currently a subject of discussion in efforts to combat the pandemic. Persistent high levels of protective antibodies in individuals who recover from COVID-19 are thought to guard against reinfection from the SARS-CoV-2 [1,2]. However, due to differences in infection conditions, treatment regimens, and individual immune status, the production and duration of protective antibodies often tend to be poles apart among recovered patients.

Several studies have confirmed that the protective antibodies, especially neutralizing antibodies (NAbs) against SARS-CoV-2, rapidly decline within a few months after recovery from the disease, risking some patients at the edge of reinfection [3–6]. Interestingly, the rate of decay and decline in protective antibodies is highly heterogeneous across individuals [7,8]. The levels of neutralizing antibodies six months, or even more, after recovery from COVID-19, remain high in a few individuals, which can help them respond rapidly to prevent reinfection. A study that investigated recovered patients who were infected with SARS-CoV-2 in the early stages, have reported that at least 90% of convalescents retained positive NAbs and SARS-CoV-2-specific T-cell responses, 6 and 12 months after the disease onset, although varying in degree [9]. In addition to the immune memory characteristics of survivors, factors during early hospitalization associated with the persistence of high levels of NAb in patients six months or longer after recovery, are still infancy and, thus, worth exploring. Given the social and economic implications of the pandemic, estimating the efficacy and duration of long-term protective antibodies after discharge from hospitals, based on early indicators, is attractive but challenging. It can also facilitate appropriate medical treatment and care in the future.

In this study, we investigate the demographic and clinical factors at admission, or the early stage of patients' hospitalization, associated with prolonged high levels of NAb against SARS-CoV-2. Data for patients that recovered from COVID-19 for at least six months were analyzed. Additionally, a model for predicting the long-term levels of NAb against COVID-19 after recovery was also constructed using the Gaussian mixture model. This model helps to guide treatment strategies and monitor responses to COVID-19 therapy, and infer long-term antibody protective efficacy from the earliest indications of hospitalization.

2. Materials and Methods

2.1. Study Population

A total of 306 individuals who recovered from COVID-19 were enrolled in this study. As described in the previous study [10], these patients were hospitalized and discharged from Tongji Hospital of the Huazhong University of Science and Technology, Wuhan, China, during 2020, as a result of laboratory-confirmed COVID-19. Given that the hospital is a local designated hospital for severe and critical illnesses, the patients had suffered moderate, severe, to critical COVID-19 infection. The disease severity was assessed at admission according to the "Chinese management guideline for COVID-19 (version 7.0)" [11]. All the convalescents included in this study were discharged from the hospital for more than six months, and none were exposed to the SARS-CoV-2 virus or suffered reinfection during the follow-up period. Simultaneously, cases included also met (1) age $\geq$ 18 years, (2) non-history of major medical or surgical conditions, such as malignant carcinoma (liver cancer, lung cancer, and so on), or organic transplantation and (3) non-psychiatric conditions, and were available for follow-up and evaluation.

Patients' categories of severity were defined as follows: moderate: patients diagnosed with COVID-19 present with fever and respiratory symptoms, and pneumonia manifestations visible via imaging. Severe: patients diagnosed with COVID-19 met any of the following criteria: (1) respiratory distress with RR $\geq$ 30 times/min; (2) peripheral oxygen saturation (SpO_2) $\leq$ 93% at rest; and (3) arterial partial pressure of oxygen (PaO_2)/fraction of inspired oxygen (FiO_2) $\leq$ 300 mmHg (1 mmHg = 0.133 kPa). Critical: meeting any of the following: (1) respiratory failure, requiring mechanical ventilation; (2) shock; (3) other organ failures, requiring intensive care unit (ICU) monitoring; or (4) death.

2.2. Data Collection

Patient data collected at baseline included: (1) demographic characteristics (age, gender, and so forth); (2) time from onset of illness to hospital admission, length of hospital stay, and disease severity; (3) clinical signs and symptoms at admission, underlying comorbidities, and treatments regimen; and (4) findings for clinical and biochemical tests at the early stage of hospitalization (initial systematic examination and comprehensive assessment of the COVID-19 patient at admission, usually within three days of hospitalization), as well as the physiological status of the patients, which were extracted from the patient's electronic medical records. Furthermore, the patient-related indicators in (1), (2), and (3) were generalized as "patient factors".

2.3. Classification of Patients

Six months after patients were discharged from the hospital, blood samples were collected from the recovered patients and assayed for NAb levels. The patients were classified into four groups based on the level of NAb against COVID-19. First, patients were classified into the NAb^{high} (high levels of NAb, n = 153) and NAb^{low} group (low levels of NAb, n = 153), based on the median level of NAb against COVID-19. Furthermore, patients in the fourth quartile (top 25% of the NAb levels) and the first quartile (the bottom 25% of the NAb levels) were further classified into the NAb^{higher} (n = 76) and NAb^{lower} group (n = 76), respectively (Figure 1).

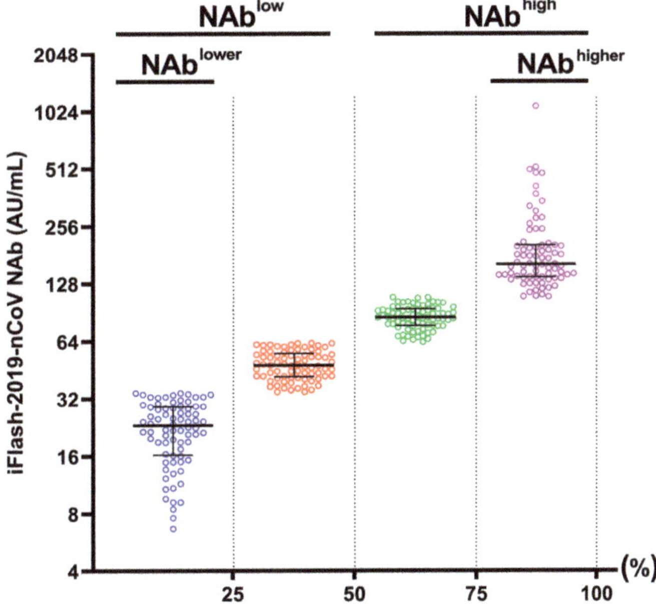

Figure 1. The level of SARS-CoV-2 NAb in 306 individuals 6 months after recovering from COVID-19. Patients were divided into NAb^{high} (median (IQR): 108.8 (85.0, 161.8) AU/mL) and NAb^{low} groups (34.9 (23.1, 48.1) AU/mL), based on the median NAb levels. In addition, 50% of individuals in the upper NAb^{high} and lower NAb^{low}, were further classified into NAb^{higher} (155.3 (116.6, 200.7) AU/mL) and NAb^{lower} groups (23.2 (16.2, 30.1) AU/mL). NAb, neutralizing antibody.

2.4. Neutralizing Antibody Assay

To evaluate the level of NAb against COVID-19, the blood samples of COVID-19 convalescents were collected and centrifuged with the assistance of a medical professional. The extracted plasma was stored at 4 °C and analyzed within 24 h. Samples that could not be analyzed within this period were stored at −80 °C and assayed within one week. The iFlash-2019-nCoV NAb kit (YHLO, Shenzhen, China, Cat: C86109) and the full-automatic chemiluminescent analyzer (iFlash 3000) were applied to assess the level of SARS-CoV-2

NAbs in plasma samples. This approach was a one-step competitive strategy chemiluminescent immunoassay (CLIA) for the quantitative detection of NAb that blocks the binding between the receptor-binding domain (RBD) and angiotensin-converting enzyme 2 (ACE2). According to the manufacturer's instructions, briefly, the plasma of samples was firstly incubated with the SARS-CoV-2 RBD antigen-coated paramagnetic microparticles. If the plasma sample contained NAb against the antigens, an antigen–antibody complex forms. The ACE2 protein acridine ester marker was then added to competitively bind the remaining RBD antigens, forming a bead-coated reaction complex. Upon introducing a magnetic field, the micro-magnetic particles were adsorbed to the reaction tube wall, but the unbound materials were washed away by the detergent. A chemiluminescent substrate was added to the immunoreactive complex, and the relative luminescence intensity (RLU) detected was inversely proportional to the number of NAbs in the plasma, which was automatically calculated and determined using the calibration curve. In particular, $\geq$10 AU/mL indicated a positive result of NAb. The superior sensitivity and specificity of this method have been validated in several studies [12–14].

2.5. Model for Predicting Levels of COVID-19 NAb

The model for predicting long-term levels of COVID-19 NAb was developed using the machine learning method of the Gaussian mixture model. After comparing the differences between the NAb^{high} group and the NAb^{low} group, factors with relative significant distinction ($p < 0.1$), namely the type of therapy received (corticosteroids therapy and intravenous immunoglobulin (IVIG) therapy); diabetes comorbidity as well as pulse oxygen saturation (SpO_2); lactate dehydrogenase (LDH) level; urea; C-reactive protein (CRP); procalcitonin (PCT); fibrinogen; and serum ferritin (SF) levels, were incorporated in the model. For the accuracy of the model, SF was not included in the model because data for many patients were missing. In addition, gender, age, and disease severity (classified as severe or above and non-severe) were also included to calibrate the model. A total of twelve candidate variables were incorporated into the model. Before modeling, continuous clinical variables were dichotomized according to the optimum cutoff value, by using the receiver operating characteristic (ROC) analysis (Table S3). The Gaussian mixture model (GMM) categorizes variables based on the hierarchical clustering of models, which features sound clustering performance and is a feasible screening method. As an unsupervised clustering, the Gaussian mixture model allows an intuitive observation of the distribution model under different combinations. Briefly, it was assumed that Gaussian distributions existed in the collected data and each distribution represented a cluster. Data points of the same distribution were first grouped together. The new probability for each data point was then assessed, followed by iterative re-classification. The highest rank in the optimal clustering would be selected after repeated training. The relationship between the various factors and levels of NAbs was assessed using univariate and multivariate regression analyses. ROC curves with AUC were constructed to assess the predictive validity of the model. Data were analyzed using the mclust package of R software (version 4.0.1).

2.6. Statistical Method

Differences between groups for categorical variables expressed as counts and percentages were analyzed using the $\chi 2$ test or Fisher's exact test, as appropriate. Continuous variables were expressed using medians and inter-quartile range (IQR), and were analyzed using the Mann–Whitney U test. Statistical significance was set at two-tailed $p < 0.05$. Data were analyzed using SPSS 26.0 (IBM Corp., Armonk, NY, USA) and R software.

3. Results

3.1. Factors at Admission Associated with NAb^{high} and NAb^{low}

Among the 306 study participants who recovered from COVID-19, 138 cases were males (45.1%). The convalescents were predominantly middle-aged and elderly persons, with a median (IQR) age of 62 (53, 68) years. NAbs were detected in the majority of the individuals (300/306, 98.0%), 6 months after their discharge. The median concentration for the NAbs was 63.1 (34.7, 108.9) AU/mL. At admission, 56.6% of patients were moderately ill, 40.7% were severely ill, and only 2.7% were critically ill. The majority of patients presented with fever (84.1%) and cough (80.5%) on admission and were appropriately treated as needed (Table S1).

Patients were divided into NAb^{high} and NAb^{low} groups, according to the median level of NAb (Figure 2e). To uncover the clinical indicators and factors associated with the persistence of high levels of NAb after hospital discharge, we compared those two groups of COVID-19 recovered patients across various parameters. Although there was a relatively high proportion of patients with severe and critical illness in the NAb^{high} group, no significant difference was found in the severity between the two groups ($p = 0.06$) (Figure 2a). It was observed that the patients in the NAb^{high} group were more likely to receive corticosteroids (38.8% vs. 22.4%, $p = 0.002$) and IVIG therapy (26.5% vs. 16.3%, $p = 0.033$) than the NAb^{low} group (Figure 2b,d). Moreover, compared to the NAb^{low} group, a substantially higher proportion of patients in the NAb^{high} group presented with underlying diabetes (25.2% vs. 12.2%, $p = 0.004$) (Figure 2c). Analysis of the physiological and biochemical test results revealed that the serum SpO_2 levels (median (IQR): 96.0 (92.0, 98.0) vs. 97.0 (94.0, 98.0)%; $p = 0.009$) at admission were relatively low in the NAb^{high} group individuals, in contrast with urea (4.8 (3.7, 6.1) vs. 3.9 (3.5, 5.4) mmol/L; $p = 0.017$); CRP (31.6 (4.0, 93.7) vs. 16.3 (2.7, 51.4) mg/L; $p = 0.027$); PCT (0.08 (0.05, 0.17) vs. 0.05 (0.03, 0.09) ng/mL; $p = 0.001$); SF (838.5 (378.2, 1533.4) vs. 478.5 (222.0, 1133.4) μg/L; $p = 0.035$); and fibrinogen (5.1 (3.8, 6.4) vs. 4.5 (3.5, 5.7) g/L; $p = 0.014$) levels, which were significantly high (Figure 2f–k). The comparison between the groups regarding other parameters is shown in Table S2.

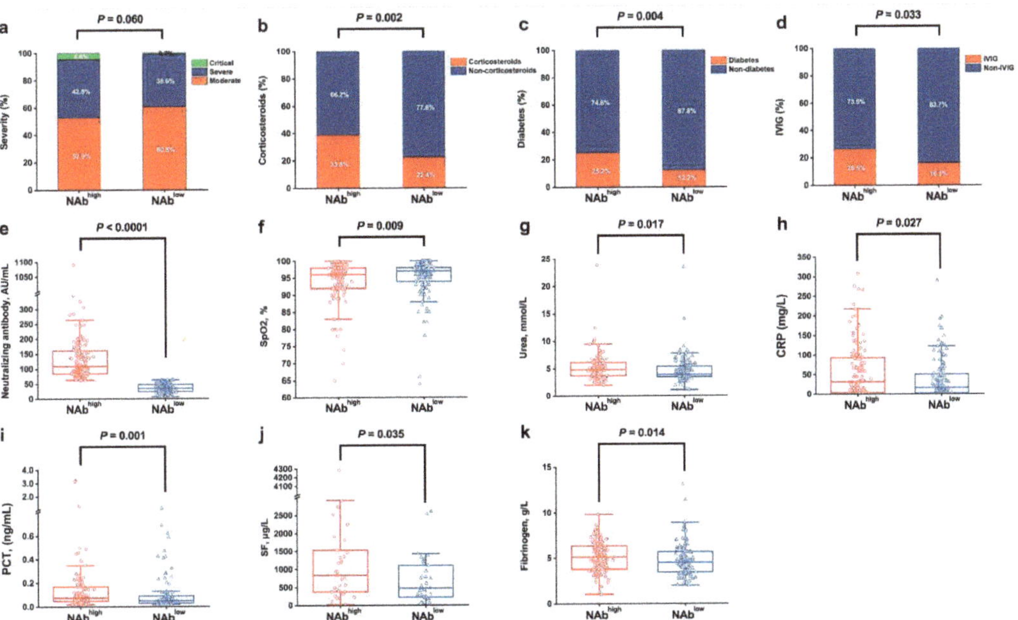

Figure 2. Various patient factors and clinical indicators between the NAb^{high} group and the NAb^{low} group at admission. (**a**), severity; (**b**), corticosteroids; (**c**), diabetes; (**d**), IVIG; (**e**), neutralizing antibody; (**f**), SpO_2; (**g**), urea; (**h**), CRP; (**i**), PCT; (**j**), SF; (**k**), fibrinogen. NAb, neutralizing antibody; IVIG, intravenous immunoglobin; SpO_2, pulse oxygen saturation; CRP, C-reactive protein; PCT, procalcitonin; and SF, serum ferritin.

3.2. *The Relationship between Long-Term Serum NAb Levels and Clinical Indicators*

To further investigate the correlation between long-term NAb levels against COVID-19 after recovery and the clinical indicators at admission, we compared the factors between individuals in the top 25% and the bottom 25% (NAb^{higher} and NAb^{lower} group) (Figure 3d). There was no significant difference in disease severity between the NAb^{higher} and NAb^{lower} groups (Figure 3a). Interestingly, we found that some clinical indicators still differed between these two groups. It was observed that compared with NAb^{lower} individuals, a higher proportion of patients in the NAb^{higher} group received corticosteroids therapy during hospitalization (45.8% vs. 21.6%, $p = 0.002$) (Figure 3b). As for comorbidity, a larger proportion of patients in the NAb^{higher} group experienced a history of diabetes at the time of admission than in the NAb^{lower} group (31.9% vs. 8.1%, $p < 0.0001$) (Figure 3c). Moreover, patients in the NAb^{higher} group displayed significantly higher levels of serum CRP (median (IQR): 24.4 (4.9, 90.3) vs. 8.5 (1.9, 32.8) mg/L; $p = 0.003$); PCT (0.07 (0.05, 0.12) vs. 0.05 (0.03, 0.07) ng/mL; $p = 0.009$); and fibrinogen (5.2 (3.8, 6.5) vs. 3.8 (3.3, 4.8) g/L; $p < 0.0001$), but lower SpO_2 (96.0 (92.0, 98.0) vs. 97.0 (95.0, 98.0) %; $p = 0.049$) levels, relative to the NAb^{lower} counterparts (Figure 3e–h).

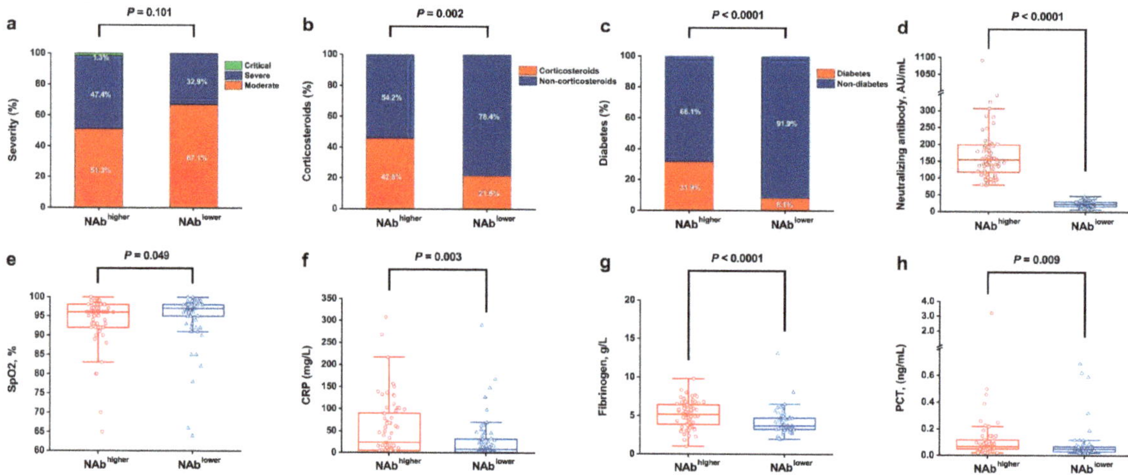

Figure 3. The difference in various indicators between the NAb^{higher} group and the NAb^{lower} group at admission. (**a**), severity; (**b**), corticosteroids; (**c**), diabetes; (**d**), neutralizing antibody; (**e**), SpO_2; (**f**), CRP; (**g**), fibrinogen; (**h**), PCT. SpO_2, pulse oxygen saturation; CRP, C-reactive protein; and PCT, procalcitonin.

These findings suggest that corticosteroids therapy and diabetes comorbidity can promote the sustained production of NAb in patients who recover from COVID-19. Additionally, the acute inflammation-related factors, such as CRP, PCT, as well as fibrinogen, and the SpO_2 levels in the initial stage of COVID-19 infection, influence the long-term production of NAb against the virus.

3.3. *Model for Predicting Long-Term NAb Levels*

For the establishment of the clinical predictive model of long-term NAb after recovery, logistic regression analyses were performed to assess the screened 12 candidate indicators. The logistic regression models for the 12 factors associated with consistently high levels of NAb had a total of 8191 formulas. Based on GMM, the 12 factors were divided into 7 clusters. After repeated training, the cluster with the highest AUC was selected for the prediction of the NAb levels of patients six months after discharge (Figure 4).

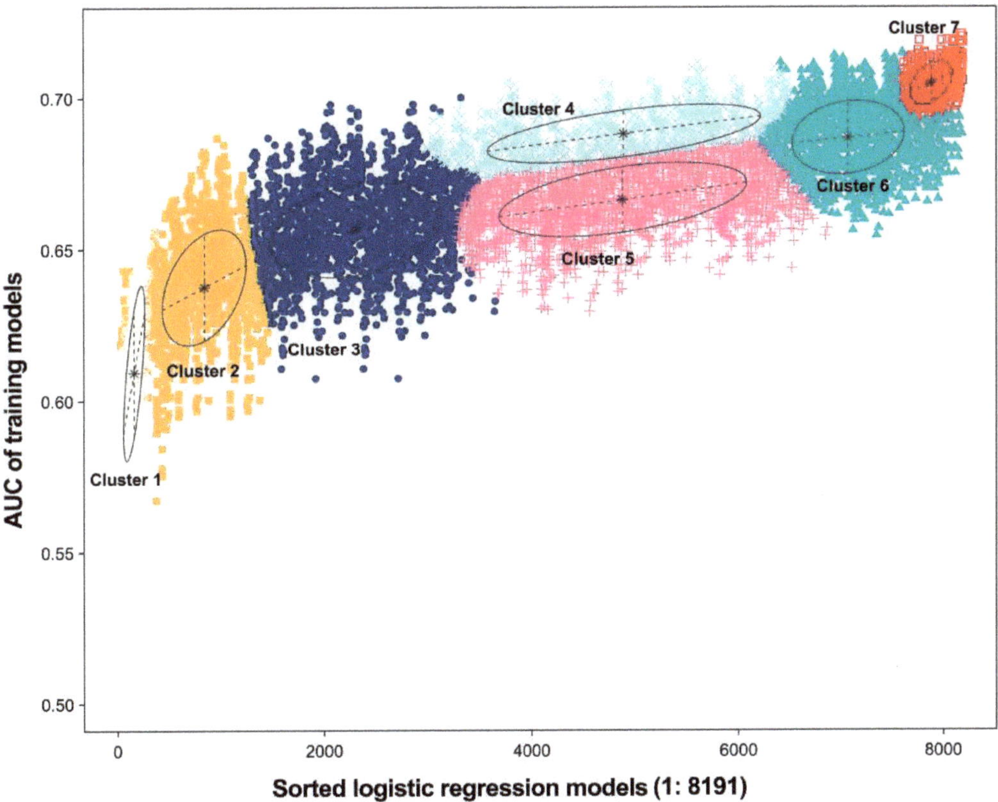

Figure 4. The logistic regression model correlated with the AUC scores, based on Gaussian mixture models for predicting the levels of NAb six months after recovery from COVID-19. There are 7 clusters of 8091 combinations; the optimal model has an average accuracy of 0.7117. Each color or shape represents a different cluster clustered by Gaussian clustering, and the horizontal coordinates represents the number of combinatorial models generated. The asterisks represent the core positions in the clusters, and the covariance matrices of the clusters bind together to form circular class clusters. The shape of the n-dimensional Gaussian distribution is determined by the covariance of each class cluster.

Overall, the developed predictive model consisted of 9 clinical indicators, including age; gender; disease severity; corticosteroids therapy; IVIG; and diabetes comorbidity, as well as the SpO_2, urea, and CRP level, which were found to influence the NAb levels six months after recovery from COVID-19. The model displayed an average accuracy of 0.7117 for the GMM classifier (Figure 4). To validate the predictive effect of the combination indicators, the ROC analysis was conducted. The predictive model had an AUC value of 0.715 (0.657–0.772), whereas its specificity and sensitivity were 72.5% and 67.3%, respectively (Figure 5).

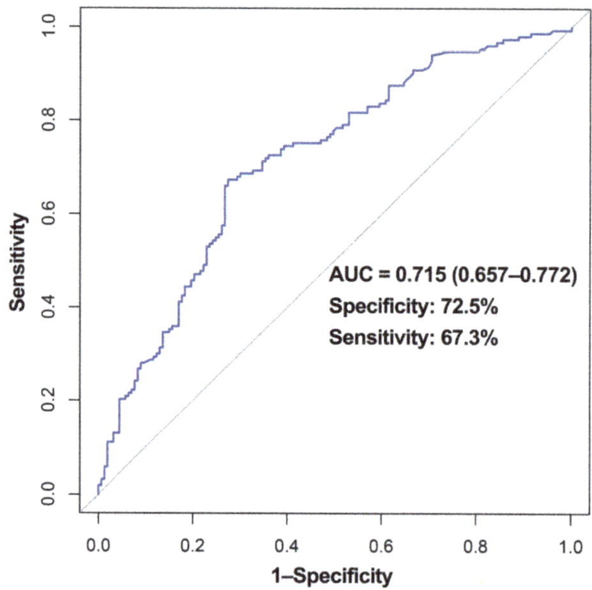

Figure 5. ROC curve for the model predicting the level of SARS-CoV-2 NAbs. The model incorporates the age and gender of the patients, the severity of the illness, therapy received (corticosteroids and IVIG), and diabetes comorbidity, as well as serum SpO_2, urea, and CRP levels. The AUC for the predictive model is 0.715 (0.657–0.772), whereas its specificity and sensitivity are 72.5% and 67.3%, respectively. IVIG, intravenous immunoglobin; SpO_2, pulse oxygen saturation; and CRP, C-reactive protein.

4. Discussion

The presence and level of SARS-CoV-2-induced neutralizing antibodies varied widely among recovered patients based on patient and treatment factors. It has been reported that serum NAb peaks within 3–5 weeks after SARS-CoV-2 infection. However, the titers and neutralizing activity decline rapidly within 1–6 months and, in some patients, the NAbs are completely undetectable several months after infection [15]. Surprisingly, in some convalescents, the titer and activity of the NAbs remain high and stable, respectively extensively beyond the follow-up period after recovery [16]. In this study, NAbs against SARS-CoV-2 were detected in 98.0% of study participants, 6 months after discharge from the hospital, despite individual differences in the levels. Earlier research also indicated consistent neutralizing activity in most subjects for as long as 6 months [17]. A study from Wuhan, China indicated that NAb concentrations in recovered patients were relatively stable for at least 9 months, regardless of whether they were symptomatic or not [8].

Existing studies suggest that NAb titers in COVID-19 survivors generally positively correlate with disease severity [18,19]. Nevertheless, in our cohort, NAb levels did not exhibit significant differences overall in patients of different disease severities (Figure S2). This discrepancy can be because all of the participants included in this study were all inpatients, who suffered from a moderate-to-critical illness. Lacking mild and asymptomatic patients for comparison, the difference in terms of disease severity was relatively small. Overall, numerous factors influence disease severity, and this complex phenomenon needs further exploration. In particular, when disease severity is similar in the infected population, the impact of clinical therapy, patient immune response, and other comorbidities during hospitalization are of concern. Diabetes comorbidity and corticosteroids therapy can contribute to high levels of NAb, six months after recovery from COVID-19. For diabetes patients, several speculations can help to explain this: (1) the disease is highly prevalent in the elderly with a natural susceptibility to COVID-19 and are likely to experience severe illness, resulting in a poor prognosis; (2) the persistent chronic inflammatory response induced by diabetes can amplify the immune response against SARS-CoV-2, resulting in more intensive and prolonged inflammatory responses that enhances the generation of memory T and B cells; and (3) the imbalance of coagulation and the fibrinolytic system in diabetic patients impairs the vascular endothelial function, which further affects the immune function and secretion of related factors. As no significant difference in disease severity was found between patients with diabetes and non-diabetes (Figure S1), the distinction of NAb levels could be due to diabetes itself.

In one study in Mexico, among 32,583 patients, diabetes was found to increase the risk of SARS-CoV-2 infection and the subsequent development of serious illness, and was related to inflammation and high mortality [20]. As a chronic inflammatory disease characterized by multiple metabolic and vascular abnormalities, diabetes promotes the production of tissue inflammation-mediated adhesion molecules and is linked with the acceleration and worsening of atherothrombosis. This increases advanced glycation end products (AGEs) and pro-inflammatory cytokines, further influencing immune responses to viral infections [21]. Pro-inflammatory cytokines, such as interleukin-6 (IL-6) and tumor necrosis factor-alpha (TNF-α) produced under viral infections, can exacerbate the severity of illness, resulting in poor prognosis, including cytokine storms.

In this study, we found that corticosteroid therapy affects the long-term production of neutralizing antibodies in different patients, but this phenomenon has not aroused concern. Corticosteroids are often widely used in the treatment of COVID-19 patients in ICU [22]. It inhibits the transcription and action of several cytokines, and modulates the proliferation, activation, differentiation, and survival of T cells and macrophages. Corticosteroids restrain the secretion of several pro-inflammatory cytokines produced by Th1 and macrophages, including IL-1β, IL-2, IL-6, and TNF-α. Patients subjected to corticosteroid therapy during hospitalization are usually in pressing need of improving clinical symptoms and oxygenation, and when the virus invades the lung epithelium, the organism is stimulated to activate specific immune cells, macrophages, and natural killer cells to produce abundant cytokines and chemokines [23]. The application of corticosteroids can alleviate chronic obstructive pulmonary disease (COPD) by modulating inflammation in the lungs [24]. Although corticosteroids can reduce the need for mechanical ventilation, they do not fully improve oxygenation in the body. Corticosteroid therapy can excessively suppress the functioning of the immune cell, thus delaying viral clearance and inducing the slow but continuous stimulation of T and B lymphocytes. Combined, these events sustain the production of numerous NAbs.

Many factors, including SpO_2, urea, CRP, PCT, SF, and fibrinogen levels, varied substantially between patients with long-term high and low levels of NAbs, and this was closely related to hypoxia, inflammation, and coagulation disorder. These indicators have been linked with the severe disease of COVID-19, in previous studies [25,26]. Low SpO_2 exacerbates pneumonic injury and the resultant hypoxemia is associated with poor clinical prognosis [27]. An immune response to COVID-19 increases the CRP, PCT, and SF, among other indicators, especially in critically ill individuals [25]. Elevated fibrinogen indicates the risk of systemic hypercoagulability and thrombotic microangiopathy in COVID-19 patients, which is inseparable from the thromboinflammation or immunothrombosis caused by COVID-19-related inflammation [28]. Blood urea levels reflect the renal function, and underlying systemic vascular and inflammatory complications. As such, it is a biomarker for inflammation-related complications [29]. Established studies also suggest that titers for NAb against SARS-CoV-2 are linked with CRP, implying that high levels of NAb can be associated with a strong inflammatory response [30,31]. In contrast, there are no significant differences in the levels of serum cytokines and chemokines between healthy individuals and asymptomatic patients. This can be because the low inflammatory responses in these individuals are insufficient to induce persistent immune responses, capable of maintaining prolonged high titers of NAb [32].

Currently, the SARS-CoV-2 is wreaking havoc across the world. However, the constant mutation of the virus strains and the rapid decay of antibodies after recovery, expose COVID-19 survivors to a plight of reinfection. An early estimate of a patient's long-term antibody levels, based on indicators at admission, can help clinicians to promptly adjust strategies during treatment and after discharge, maximizing the maintenance of high levels of long-term protective antibodies and reducing the risk of reinfection. Applying this predictive model at an early assessment, provides a clinical reference for treatment decisions, inpatient care, and individualized programs for COVID-19 patients, effectively enabling target measures to enhance post-discharge antiviral resistance and immune protection.

Regarding the limitations, firstly, we were not able to monitor and compare ongoing changes in antibody levels in convalescents over multiple periods. Secondly, given the retrospective nature of the research, the test results of clinical indicators for some patients were incomplete when information was extracted. Lastly, the established prediction model requires further validation in a subsequently larger clinical population.

5. Conclusions

In conclusion, several factors at admission can contribute to a high level of NAbs in patients, six months after discharge. We found that diabetes comorbidity and corticosteroids therapy, as well as the SpO_2, CRP, PCT, and fibrinogen levels, affect the prolonged production of NAbs against SARS-CoV-2 in individuals who recover from COVID-19. In addition, we constructed a predictive model for sustained NAb levels in convalescents. The findings of this study can provide some valuable guidance for the treatment and monitoring of patients in clinical recovery.

Supplementary Materials: The following are available online at https://www.mdpi.com/article/10.3390/v14010080/s1, Table S1: comparison of demographic and clinical indicators of admission between the NAb^{high} and NAb^{low} groups; Table S2: comparison of demographic and clinical indicators of admission between the NAb^{higher} and NAb^{lower} groups; Table S3: cut-off values for the dichotomous categories in the ROC analysis of the continuous variables before GMM model; Figure S1: distribution of disease severity in diabetic and non-diabetic patients among 306 patients recovered from COVID-19; and Figure S2: NAb levels in patients with different disease severity.

Author Contributions: Conceptualization, methodology, and supervision, T.S., D.X. and S.X.; validation and investigation, X.L., L.P., Y.Y. and Y.Z.; formal analysis, X.L. and L.P.; visualization and writing—original draft preparation, X.L.; writing—review and editing, all authors. All authors have read and agreed to the published version of the manuscript.

Funding: The study was supported by the Beijing Municipal Natural Science Foundation (M21021) and the National Natural Science Foundation of China (82072277).

Institutional Review Board Statement: The protocol of this study was approved by the Institutional Review Board of Tongji Hospital, Tongji Medical College, Huazhong University of Science and Technology (ethics approval number: TJ-IRB20210115).

Informed Consent Statement: Due to the retrospective nature of the study, the ethics committee of the designated hospital waived the requirement for informed consent.

Data Availability Statement: The data used to support the findings in this study are available from the corresponding authors upon request.

Acknowledgments: The authors thank the individuals recovered from the COVID-19 who participated in this study, as well as the great medical workers fighting the outbreak.

Conflicts of Interest: The authors declare no conflict of interest.

References

1. Khoshkam, Z.; Aftabi, Y.; Stenvinkel, P.; Paige Lawrence, B.; Rezaei, M.H.; Ichihara, G.; Fereidouni, S. Recovery scenario and immunity in COVID-19 disease: A new strategy to predict the potential of reinfection. *J. Adv. Res.* **2021**, *31*, 49–60. [CrossRef]
2. Cromer, D.; Juno, J.A.; Khoury, D.; Reynaldi, A.; Wheatley, A.K.; Kent, S.J.; Davenport, M.P. Prospects for durable immune control of SARS-CoV-2 and prevention of reinfection. *Nat. Rev. Immunol.* **2021**, *21*, 395–404. [CrossRef] [PubMed]

3. Muecksch, F.; Wise, H.; Batchelor, B.; Squires, M.; Semple, E.; Richardson, C.; McGuire, J.; Clearly, S.; Furrie, E.; Greig, N.; et al. Longitudinal Serological Analysis and Neutralizing Antibody Levels in Coronavirus Disease 2019 Convalescent Patients. *J. Infect. Dis.* **2021**, *223*, 389–398. [CrossRef] [PubMed]
4. Zuo, J.; Dowell, A.C.; Pearce, H.; Verma, K.; Long, H.M.; Begum, J.; Aiano, F.; Amin-Chowdhury, Z.; Hoschler, K.; Brooks, T.; et al. Robust SARS-CoV-2-specific T cell immunity is maintained at 6 months following primary infection. *Nat. Immunol.* **2021**, *22*, 620–626. [CrossRef] [PubMed]
5. Röltgen, K.; Powell, A.E.; Wirz, O.F.; Stevens, B.A.; Hogan, C.A.; Najeeb, J.; Hunter, M.; Wang, H.; Sahoo, M.K.; Huang, C.; et al. Defining the features and duration of antibody responses to SARS-CoV-2 infection associated with disease severity and outcome. *Sci. Immunol.* **2020**, *5*, eabe0240. [CrossRef]
6. Ibarrondo, F.J.; Fulcher, J.A.; Goodman-Meza, D.; Elliott, J.; Hofmann, C.; Hausner, M.A.; Ferbas, K.G.; Tobin, N.H.; Aldrovandi, G.M.; Yang, O.O. Rapid Decay of Anti-SARS-CoV-2 Antibodies in Persons with Mild COVID-19. *N. Engl. J. Med.* **2020**, *383*, 1085–1087. [CrossRef]
7. Chia, W.N.; Zhu, F.; Ong, S.W.X.; Young, B.E.; Fong, S.-W.; Le Bert, N.; Tan, C.W.; Tiu, C.; Zhang, J.; Tan, S.Y.; et al. Dynamics of SARS-CoV-2 neutralising antibody responses and duration of immunity: A longitudinal study. *Lancet Microbe* **2021**, *2*, e240–e249. [CrossRef]
8. He, Z.; Ren, L.; Yang, J.; Guo, L.; Feng, L.; Ma, C.; Wang, X.; Leng, Z.; Tong, X.; Zhou, W.; et al. Seroprevalence and humoral immune durability of anti-SARS-CoV-2 antibodies in Wuhan, China: A longitudinal, population-level, cross-sectional study. *Lancet* **2021**, *397*, 1075–1084. [CrossRef]
9. Zhang, J.; Lin, H.; Ye, B.; Zhao, M.; Zhan, J.; Dong, S.; Guo, Y.; Zhao, Y.; Li, M.; Liu, S.; et al. One-year sustained cellular and humoral immunities of COVID-19 convalescents. *Clin. Infect. Dis.* **2021**, ciab884. [CrossRef]
10. Chen, H.; Li, X.; Marmar, T.; Xu, Q.; Tu, J.; Li, T.; Han, J.; Xu, D.; Shen, T. Cardiac Troponin I association with critical illness and death risk in 726 seriously ill COVID-19 patients: A retrospective cohort study. *Int. J. Med. Sci.* **2021**, *18*, 1474–1483. [CrossRef]
11. NHC. National Health Commission of China. Chinese Management Guideline for COVID-19 (version 7.0). Available online: http://www.nhc.gov.cn/yzygj/s7653p/202003/46c9294a7dfe4cef80dc7f5912eb1989.shtml (accessed on 25 March 2021).
12. Tenbusch, M.; Schumacher, S.; Vogel, E.; Priller, A.; Held, J.; Steininger, P.; Beileke, S.; Irrgang, P.; Brockhoff, R.; Salmanton-García, J.; et al. Heterologous prime–boost vaccination with ChAdOx1 nCoV-19 and BNT162b2. *Lancet Infect. Dis.* **2021**, *21*, 1212–1213. [CrossRef]
13. Pan, Y.; Jiang, X.; Yang, L.; Chen, L.; Zeng, X.; Liu, G.; Tang, Y.; Qian, C.; Wang, X.; Cheng, F.; et al. SARS-CoV-2-specific immune response in COVID-19 convalescent individuals. *Signal Transduct. Target. Ther.* **2021**, *6*, 256. [CrossRef]
14. Favresse, J.; Gillot, C.; Di Chiaro, L.; Eucher, C.; Elsen, M.; Van Eeckhoudt, S.; David, C.; Morimont, L.; Dogné, J.-M.; Douxfils, J. Neutralizing Antibodies in COVID-19 Patients and Vaccine Recipients after Two Doses of BNT162b2. *Viruses* **2021**, *13*, 1364. [CrossRef]
15. Gaebler, C.; Wang, Z.; Lorenzi, J.C.C.; Muecksch, F.; Finkin, S.; Tokuyama, M.; Cho, A.; Jankovic, M.; Schaefer-Babajew, D.; Oliveira, T.Y.; et al. Evolution of antibody immunity to SARS-CoV-2. *Nature* **2021**, *591*, 639–644. [CrossRef] [PubMed]
16. Shi, D.; Weng, T.; Wu, J.; Dai, C.; Luo, R.; Chen, K.; Zhu, M.; Lu, X.; Cheng, L.; Chen, Q.; et al. Dynamic Characteristic Analysis of Antibodies in Patients with COVID-19: A 13-Month Study. *Front. Immunol.* **2021**, *12*, 708184. [CrossRef]
17. Figueiredo-Campos, P.; Blankenhaus, B.; Mota, C.; Gomes, A.; Serrano, M.; Ariotti, S.; Costa, C.; Nunes-Cabaço, H.; Mendes, A.M.; Gaspar, P.; et al. Seroprevalence of anti-SARS-CoV-2 antibodies in COVID-19 patients and healthy volunteers up to 6 months post disease onset. *Eur. J. Immunol.* **2020**, *50*, 2025–2040. [CrossRef] [PubMed]
18. Chen, X.; Pan, Z.; Yue, S.; Yu, F.; Zhang, J.; Yang, Y.; Li, R.; Liu, B.; Yang, X.; Gao, L.; et al. Disease severity dictates SARS-CoV-2-specific neutralizing antibody responses in COVID-19. *Signal Transduct. Target. Ther.* **2020**, *5*, 180. [CrossRef]
19. Legros, V.; Denolly, S.; Vogrig, M.; Boson, B.; Siret, E.; Rigaill, J.; Pillet, S.; Grattard, F.; Gonzalo, S.; Verhoeven, P.; et al. A longitudinal study of SARS-CoV-2-infected patients reveals a high correlation between neutralizing antibodies and COVID-19 severity. *Cell. Mol. Immunol.* **2021**, *18*, 318–327. [CrossRef] [PubMed]
20. Wang, Z.; Du, Z.; Zhu, F. Glycosylated hemoglobin is associated with systemic inflammation, hypercoagulability, and prognosis of COVID-19 patients. *Diabetes Res. Clin. Pract.* **2020**, *164*, 108214. [CrossRef]
21. Hussain, A.; Bhowmik, B.; do Vale Moreira, N.C. COVID-19 and diabetes: Knowledge in progress. *Diabetes Res. Clin. Pract.* **2020**, *162*, 108142. [CrossRef]
22. Tang, Y.; Liu, J.; Zhang, D.; Xu, Z.; Ji, J.; Wen, C. Cytokine Storm in COVID-19: The Current Evidence and Treatment Strategies. *Front. Immunol.* **2020**, *11*, 1708. [CrossRef]
23. Yang, J.W.; Yang, L.; Luo, R.G.; Xu, J.F. Corticosteroid administration for viral pneumonia: COVID-19 and beyond. *Clin. Microbiol. Infect.* **2020**, *26*, 1171–1177. [CrossRef] [PubMed]
24. Halpin, D.M.G.; Criner, G.J.; Papi, A.; Singh, D.; Anzueto, A.; Martinez, F.J.; Agusti, A.A.; Vogelmeier, C.F. Global Initiative for the Diagnosis, Management, and Prevention of Chronic Obstructive Lung Disease. The 2020 GOLD Science Committee Report on COVID-19 and Chronic Obstructive Pulmonary Disease. *Am. J. Respir. Crit. Care Med.* **2021**, *203*, 24–36. [CrossRef]
25. Gao, Y.-D.; Ding, M.; Dong, X.; Zhang, J.-J.; Kursat Azkur, A.; Azkur, D.; Gan, H.; Sun, Y.-L.; Fu, W.; Li, W.; et al. Risk factors for severe and critically ill COVID-19 patients: A review. *Allergy* **2021**, *76*, 428–455. [CrossRef] [PubMed]
26. Li, X.; Marmar, T.; Xu, Q.; Tu, J.; Yin, Y.; Tao, Q.; Chen, H.; Shen, T.; Xu, D. Predictive indicators of severe COVID-19 independent of comorbidities and advanced age: A nested case-control study. *Epidemiol. Infect.* **2020**, *148*, e255. [CrossRef]

27. Gallo Marin, B.; Aghagoli, G.; Lavine, K.; Yang, L.; Siff, E.J.; Chiang, S.S.; Salazar-Mather, T.P.; Dumenco, L.; Savaria, M.C.; Aung, S.N.; et al. Predictors of COVID-19 severity: A literature review. *Rev. Med. Virol.* **2021**, *31*, 1–10. [CrossRef]
28. Connors, J.M.; Levy, J.H. COVID-19 and its implications for thrombosis and anticoagulation. *Blood* **2020**, *135*, 2033–2040. [CrossRef]
29. Gong, J.; Ou, J.; Qiu, X.; Jie, Y.; Chen, Y.; Yuan, L.; Cao, J.; Tan, M.; Xu, W.; Zheng, F.; et al. A Tool for Early Prediction of Severe Coronavirus Disease 2019 (COVID-19): A Multicenter Study Using the Risk Nomogram in Wuhan and Guangdong, China. *Clin. Infect. Dis.* **2020**, *71*, 833–840. [CrossRef]
30. Wu, F.; Liu, M.; Wang, A.; Lu, L.; Wang, Q.; Gu, C.; Chen, J.; Wu, Y.; Xia, S.; Ling, Y.; et al. Evaluating the Association of Clinical Characteristics with Neutralizing Antibody Levels in Patients Who Have Recovered from Mild COVID-19 in Shanghai, China. *JAMA Intern. Med.* **2020**, *180*, 1356–1362. [CrossRef]
31. Chen, W.; Zhang, J.; Qin, X.; Wang, W.; Xu, M.; Wang, L.-F.; Xu, C.; Tang, S.; Liu, P.; Zhang, L.; et al. SARS-CoV-2 neutralizing antibody levels are correlated with severity of COVID-19 pneumonia. *Biomed. Pharmacother.* **2020**, *130*, 110629. [CrossRef] [PubMed]
32. Long, Q.-X.; Tang, X.-J.; Shi, Q.-L.; Li, Q.; Deng, H.-J.; Yuan, J.; Hu, J.-L.; Xu, W.; Zhang, Y.; Lv, F.-J.; et al. Clinical and immunological assessment of asymptomatic SARS-CoV-2 infections. *Nat. Med.* **2020**, *26*, 1200–1204. [CrossRef] [PubMed]

 viruses MDPI

Article

Distinct Outcomes in COVID-19 Patients with Positive or Negative RT-PCR Test

Maria Clara Saad Menezes [1,*], Diego Vinicius Santinelli Pestana [1], Juliana Carvalho Ferreira [2], Carlos Roberto Ribeiro de Carvalho [2], Marcelo Consorti Felix [2], Izabel Oliva Marcilio [2], Katia Regina da Silva [2], Vilson Cobello Junior [2], Julio Flavio Marchini [2], Julio Cesar Alencar [2], Luz Marina Gomez Gomez [1], Denis Deratani Mauá [3], Heraldo Possolo Souza [1], Emergency USP COVID-19 Group [1,†] and HCFMUSP COVID-19 Study Group [2,‡]

1 Emergency Medicine Department, Faculdade de Medicina da Universidade de São Paulo, São Paulo 01246-903, Brazil; diego.pestana@fm.usp.br (D.V.S.P.); lgomez928@gmail.com (L.M.G.G.); heraldo.possolo@fm.usp.br (H.P.S.)

2 Hospital das Clínicas, Faculdade de Medicina da Universidade de São Paulo, São Paulo 01246-9031, Brazil; juliana.ferreira@hc.fm.usp.br (J.C.F.); carlos.carvalho@hc.fm.usp.br (C.R.R.d.C.); marcelo.felix@hc.fm.usp.br (M.C.F.); izamarcilio@gmail.com (I.O.M.); katia.research@gmail.com (K.R.d.S.); vilson.cobello@hc.fm.usp.br (V.C.J.); julio.marchini@hc.fm.usp.br (J.F.M.); julio.alencar@hc.fm.usp.br (J.C.A.)

3 Institute of Mathematics and Statistics, Universidade de São Paulo, São Paulo 05508-090, Brazil; denis.maua@usp.br

* Correspondence: mclarasaad22@gmail.com

† Emergency USP COVID-19 Group: Alicia Dudy Müller Veiga, Arthur Petrillo Bellintani, Helena Souza Pinto, Isabela Argollo Ferreira, Joana Margarida Alves Mota, Luca Suzano Baptista, Marcelo de Oliveira Silva, Mário Turolla Ribeiro, Matheus Nakao de Sousa, Matheus Saldanha Rodrigues Duarte, Matheus Zanelatto Junqueira, Mônica Yhasmin de Lima Redondo, Pedro Gaspar dos Santos, Rafael Berenguer Luna, Tales Cabral Monsalvarga, Thalissa Ferreira, Vitor Macedo Brito Medeiros, Willaby Serafim Cassa Ferreira.

‡ HCFMUSP COVID-19 Study Group: Tarcisio E.P. Barros-Filho, Eloisa Bonfa, Edivaldo M. Utiyama, Aluisio C. Segurado, Beatriz Perondi, Anna Miethke-Morais, Amanda C. Montal, Leila Harima, Solange R.G. Fusco, Marjorie F. Silva, Marcelo C. Rocha, Izabel Marcilio, Izabel Cristina Rios, Fabiane Yumi Ogihara Kawano, Maria Amélia de Jesus, Esper Kallas, Maria Cristina Peres Braido Francisco, Carolina Mendes do Carmo, Clarice Tanaka, Maura Salaroli Oliveira, Thaís Guimarães, Carolina dos Santos Lázari, Marcello M.C. Magri, Anna Sara Shafferman Levin, Alberto José da Silva Duarte, Ester Sabino.

Citation: Saad Menezes, M.C.; Santinelli Pestana, D.V.; Ferreira, J.C.; Ribeiro de Carvalho, C.R.; Felix, M.C.; Marcilio, I.O.; da Silva, K.R.; Junior, V.C.; Marchini, J.F.; Alencar, J.C.; et al. Distinct Outcomes in COVID-19 Patients with Positive or Negative RT-PCR Test. *Viruses* **2022**, *14*, 175. https://doi.org/10.3390/v14020175

Academic Editors: Burtram C. Fielding and Georgia Schäfer

Received: 7 December 2021
Accepted: 15 January 2022
Published: 18 January 2022

Publisher's Note: MDPI stays neutral with regard to jurisdictional claims in published maps and institutional affiliations.

Abstract: Identification of the SARS-CoV-2 virus by RT-PCR from a nasopharyngeal swab sample is a common test for diagnosing COVID-19. However, some patients present clinical, laboratorial, and radiological evidence of COVID-19 infection with negative RT-PCR result(s). Thus, we assessed whether positive results were associated with intubation and mortality. This study was conducted in a Brazilian tertiary hospital from March to August of 2020. All patients had clinical, laboratory, and radiological diagnosis of COVID-19. They were divided into two groups: positive (+) RT-PCR group, with 2292 participants, and negative (−) RT-PCR group, with 706 participants. Patients with negative RT-PCR testing and an alternative most probable diagnosis were excluded from the study. The RT-PCR(+) group presented increased risk of intensive care unit (ICU) admission, mechanical ventilation, length of hospital stay, and 28-day mortality, when compared to the RT-PCR(−) group. A positive SARS-CoV-2 RT-PCR result was independently associated with intubation and 28 day in-hospital mortality. Accordingly, we concluded that patients with a COVID-19 diagnosis based on clinical data, despite a negative RT-PCR test from nasopharyngeal samples, presented more favorable outcomes than patients with positive RT-PCR test(s).

Keywords: SARS-CoV-2; COVID-19 testing; COVID-19; hospital mortality; intubation

1. Introduction

COVID-19 patients usually present systemic and respiratory symptoms, such as fever, cough, and shortness of breath [1]. Radiological exams may show different degrees of lung involvement, and laboratory tests may show increased inflammatory markers [1–3]. None of these factors are specific, and the diagnosis is obtained only when the SARS-CoV-2 virus is detected in the airways [4].

During the initial times of the pandemic, there were concerns regarding the accuracy of the tests used [5]. A meta-analysis published in May 2020 found that the accuracy of the RT-PCR test for coronavirus diagnosis can change according to the prevalence of COVID-19. With a prevalence of 50%, common among health professionals with respiratory symptoms, a post-test probability of 96% was found. With a prevalence of 20%, the post-test probability was 84%. With a prevalence of 5%, they found a 55% post-test probability [6]. More recently, another meta-analysis showed that the test's accuracy has improved over the year; however, a marked heterogeneity in the proportion of false-negative RT-PCR results amongst different tests is still maintained [7]. The heterogeneity is largely unexplained. There are several reasons that can underlie this heterogeneity. Researchers have suggested that these failures in SARS-CoV-2 detection are related to multiple preanalytical and analytical factors, such as lack of standardization for specimen collection, delays, or poor storage conditions before arrival in the laboratory, the use of inadequately validated assays, contamination during the procedure, insufficient viral specimens and load, the incubation period of the disease, and the presence of mutations that escape detection [7].

Nevertheless, even with improvements in virus detection, there is a group of patients diagnosed with COVID-19 based solely on clinical criteria because they do not present a positive RT-PCR test [8]. In this group of patients, the differential diagnosis for COVID-19 should be ruled out. This group of differential diseases includes mainly respiratory diseases, infections of other origins, cardiovascular, oncological, gastrointestinal, urogenital, and neurological diseases [9].

Clinical criteria for diagnosing COVID-19 include the initial assessment of related symptoms and exposure history [2], coupled with typical laboratory findings (lymphopenia, thrombocytopenia, and elevated liver enzymes, lactate dehydrogenase, inflammatory markers, and D-dimer) [3], and characteristic images at the lung CT scan (multiple bilateral ground-glass opacities in the peripheral lower lung zones) [10].

These presumptive diagnostic criteria have been reported to be more sensitive than the RT-PCR, providing a positive result at earlier stages of the disease [11]. Currently, both methods are described as complementary and should be used in conjunction when available.

Notwithstanding, it is unclear whether patients with a presumptive diagnosis of COVID-19 evolve similarly to patients who were diagnosed by the RT-PCR test. Therefore, in this study, we explored the specificities of the aforementioned group by comparing 2292 COVID-19 hospitalized patients confirmed by SARS-CoV-2 RT-PCR (COVID-19 RT-PCR(+) group) with 706 COVID-19 hospitalized patients diagnosed by presumptive clinical criteria with negative SARS-CoV-2 RT-PCR results (COVID-19 RT-PCR(−) group).

2. Materials and Methods

We conducted a retrospective unicentric cohort study from March to August 2020 at the Hospital das Clínicas da Universidade de São Paulo (HC-FMUSP), a 2200-bed urban, academic medical center comprising five institutes and two auxiliary hospitals. During the pandemic, the HC-FMUSP has been designated for the reception and care of patients with severe COVID-19. All patients were evaluated and treated according to the Institution's protocol that consisted and still consists of supportive care, including supplemental oxygen and mechanical ventilatory support when indicated. By the end of July 2020, after the preliminary report from the RECOVERY Collaborative Group, patients who were receiving either invasive mechanical ventilation or oxygen alone were also treated with corticosteroids both in the COVID-19 RT-PCR(+) group and COVID-19 RT-PCR(−) group. Therefore, there were no differences in the treatment provided for both groups. There were no instances during the time-period studied when mechanical ventilation or other treatment modalities were unavailable for patients who might have needed them.

In this research study, the COVID-19 RT-PCR(+) group is formed by patients with a high clinical suspicion of COVID-19 and a positive RT-PCR test for the SARS-CoV-2 whereas the COVID-19 RT-PCR(−) group is formed by patients with a high clinical suspicion of COVID-19, judged by 2 experienced attending physicians after ruling out differential diagnosis, and at least two negative RT-PCR tests for the SARS-CoV-2.

2.1. COVID-19 Diagnostics

Individuals with clinical suspicion of severe COVID-19 were referred to our tertiary hospital from other healthcare institutions in our city when the physician had a suspicion for COVID-19. All subjects had some sort of respiratory symptoms (cough or dyspnea) and diffuse infiltrates on X-rays.

Upon arrival, these patients were classified according to the RT-PCR test results for SARS-CoV-2 detection. Patients that arrived with a previous nasopharyngeal swab RT-PCR test result performed at the origin hospital were immediately included in the COVID-19 RT-PCR(+) group.

Patients with no previous test or a negative test were then submitted to the RT-PCR test at hospital admission. Patients with positive RT-PCR results were then allocated to the COVID-19 RT-PCR(+) group.

Patients with negative tests were then submitted to lung CT scans, laboratory tests, and another nasopharyngeal swab collection. When a diagnosis other than COVID-19 was made, the subject was excluded from our study. If the patient had more than 7 days of COVID-19 symptoms on hospital admission, COVID-19 serology was also performed. During the inpatient stay, patients that had at least two negative RT-PCR tests but maintained high clinical suspicion for COVID-19 also underwent serological testing after, at least, 7 days of symptoms. Unfortunately, due to a shortage of resources, serologic COVID-19 testing could not be performed in all eligible patients. COVID-19 serology at discharge was not performed. In every serological testing occasion, both IgM and IgG were tested, and a positive result in either one was considered enough to consider that patients had positive serology status.

At discharge, patients with no positive SARS-CoV-2 RT-PCR result but who were considered by sequential attending physicians (at least 2) with clinical and radiological suspicion of COVID-19, after ruling out differential diagnoses, were allocated to the COVID-19 RT-PCR(−) group. Scheme 1 illustrates the institutional testing protocol and the distribution of participants. Of note, CT scans were reported by specialist radiologists, and the results were freely available to those who adjudicated on the presence or absence of COVID-19.

Therefore, the RT-PCR(+) group was composed of patients with clinical, laboratory, and radiological findings of COVID-19 and a positive nasopharyngeal swab for SARS-CoV-2 at some point of the disease evolution. The RT-PCR(−) group was composed of patients with clinical, laboratory, and radiological findings of COVID-19, other differential diagnoses excluded, and no positive RT-PCR test for SARS-CoV-2 during their hospital treatment.

Patients with a differential diagnosis more probable than COVID-19 were excluded as well as patients younger than 18 years of age. Before hospital discharge, all patients were assessed for the presence of antibodies against SARS-CoV-2 in circulating blood. Routine screening for other respiratory viruses was undertaken whenever there was uncertainty regarding the COVID-19 diagnosis. Patients with a positive result for another virus were excluded from this study.

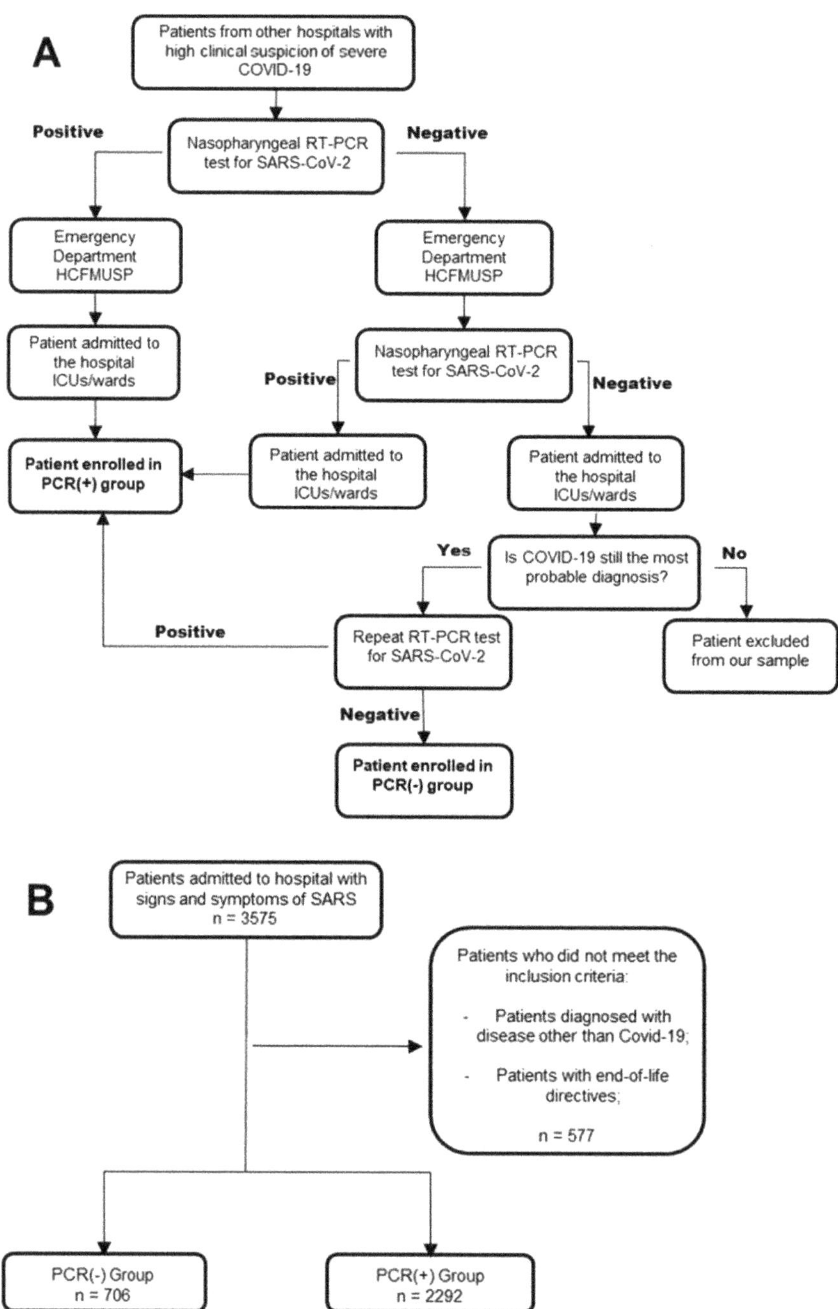

Scheme 1. (**A**) Institutional protocol for SARS-CoV-2 RT-PCR collection. (**B**) Distribution of participants.

2.2. Data Collection

We designed an extensive database to characterize all hospital admissions for COVID-19 from March to August 2020. Datasets (demographics, comorbidities, clinical conditions, treatment, laboratory tests, and outcomes) derived from the electronic health record were directly imported to the Research Electronic Data Capture (REDCap) [12] system hosted at Hospital das Clínicas da Faculdade de Medicina da Universidade de São Paulo. These datasets were submitted to a rigorous process of harmonization and consolidation before being imported to REDCap. In addition, a task force of researchers reviewed the medical records to ensure data completeness and quality.

2.3. RT-PCR Assessment

The patient's nasopharyngeal or oropharyngeal secretions, or both, were collected via swab. An automated separation kit was used to extract and purify viral RNA from samples using magnetic microparticles (Abbott mSample Preparation Kit RNA, Des Plaines, IL, USA). The reverse transcription, amplification, and real-time detection protocols were validated by our institutional Laboratory Division (accredited by the College of American Pathologists). First- and second-line screening tools, E gene assay and N gene assay, respectively, were used for confirmation, as described by Corman et al. [13]. Extraction and amplification were internally controlled in every sample using the endogenous gene RNAseP, besides positive and negative controls in all batches. The analytical sensitivity was 40 RNA copies/mL, and specificity was 100% even in samples containing other respiratory viruses. Of note, some patients were transferred from smaller facilities to our institution and were tested for SARS-CoV-2 infection before admission in our hospital; tracking the exact kit and reagents used for testing this subset of patients was not feasible.

The Abbott mSample Preparation Kit RNA has been authorized by the United States of America Food and Drugs Administration (FDA) as of July 2020 and is currently in accordance with the latest recommendations made by the Center of Disease Control and Prevention (CDC) regarding the use of nucleic acid amplification tests for diagnosing SARS-CoV-2.

2.4. Statistical Analysis

Baseline characteristics and outcomes of hospitalized patients in the COVID-19 RT-PCR(+) group and COVID-19 RT-PCR(−) group were compared using the Mann–Whitney test. Continuous variables were represented with medians, and interquartile ranges and categorical variables were represented as proportions. A 2-sided p value of ≤ 0.05 was used to designate statistical significance.

The survival probability and probability of not being submitted to intubation in the RT-PCR(+) group and RT-PCR(−) were calculated using the Kaplan–Meier method. Hazard ratio of the risk variables were estimated by fitting a Cox proportional hazards model. Afterwards, baseline factors that were associated with 28-day in-hospital mortality and intubation were identified using univariate logistic regression models. All variables that were statistically significantly associated with each outcome were then entered into separate multivariate logistic regression models. Adjusted odds ratios of mortality and intubation were calculated for each of these variables with 95% confidence intervals (CIs). Analyses were conducted using the Python frameworks statsmodels (v. 0.12.2), scikit-learn (v. 0.24.2) and lifelines (v. 0.26). Eventual missing data were set to be managed according to their nature. Values missed completely at random were addressed using complete case analysis; values missing at random were addressed using multiple imputation methods; values likely to be missing not at random will be re-evaluated by authors.

3. Results

3.1. Patients' Characteristics and Presentation

A total of 2998 patients were included in this study. Groups were divided according to their SARS-CoV-2 RT-PCR testing status, as described above. A total of seven hundred and six (23.5%) patients were allocated to the COVID-19 RT-PCR(−) group, and two thousand two hundred and ninety-two (76.5%) patients were allocated to the COVID-19 RT-PCR(+)

group. A total of five hundred and seventy-seven patients were excluded from the study for having a more likely alternative diagnosis other than COVID-19.

Demographic variables were similar between the two experimental groups, as shown in Table 1. Regarding comorbidities, a higher prevalence of diabetes, cardiovascular diseases, chronic kidney disease, peripheral vascular disease, and cancer was observed in the RT-PCR(+) group. Regarding clinical presentation at hospital admission, patients in the RT-PCR(−) group had a delay in seeking medical attention when compared to patients in the RT-PCR(+) group (Table 2). The median time after initial symptoms was eight days (IQR 5–11) in the former group, compared to seven days (IQR 4–10) in the latter ($p < 0.001$, 95% CI).

Table 1. Comparison of demographic characteristics and frequency of comorbidities in the COVID-19 RT-PCR(+) group and COVID-19 RT-PCR(−) group.

Characteristics and Comorbidities (*n* = 2998)	COVID-19 PCR Positive *n* = 2292	COVID-19 PCR Negative *n* = 706	*p*-Value *,1
Age, years	61 (48–71) [2292]	60 (47–72) [706]	0.262
Male Sex	1281/2292 (55.9%)	367/706 (52.0%)	0.068
Body mass index **,2	26 (23–32) [1901]	26 (23–33) [490]	0.710
Previous Diseases			
Hypertension **	1325/2291 (57.8%)	402/704 (57.1%)	0.731
Diabetes **	873/2290 (38.1%)	224/706 (31.7%)	0.002
Cardiovascular disease **	464/2289 (20.3%)	115/702 (16.4%)	0.023
Former or current smoker **	488/2283 (21.4%)	164/702 (23.4%)	0.265
Current smoker **	147/2287 (6.4%)	42/705 (6.0%)	0.654
Chronic kidney disease **	227/2289 (9.9%)	51/705 (7.2%)	0.032
Active cancer **	279/2128 (13.1%)	53/605 (8.8%)	0.004
Cerebrovascular disease	173/2292 (7.5%)	48/706 (6.8%)	0.505
Arrhythmia **	116/1757 (6.6%)	37/492 (7.5%)	0.475
Peripheral Vascular Disease **	107/1707 (6.3%)	14/460 (3.0%)	0.008
Chronic obstructive pulmonary disease **	151/2291 (6.6%)	44/705 (6.2%)	0.742
Asthma **	91/2292 (4.0%)	31/705 (4.4%)	0.616
End-stage renal disease **	90/2290 (3.9%)	20/705 (2.8%)	0.177
Hematologic malignancy **	80/1743 (4.6%)	10/474 (2.1%)	0.015
Rheumatologic disease **	62/2291 (2.7%)	23/706 (3.3%)	0.440
Liver disease **	74/2291 (3.2%)	17/706 (2.4%)	0.266
Human immunodeficiency virus infection	20/2292 (0.9%)	12/706 (1.7%)	0.062

1.* *p* values were calculated using the nonparametric Mann–Whitney U command in Python (v 0.24.1), version 15.0, that tests for trend across ordered groups. Comparing COVID-19 RT-PCR positive and negative groups. 2.** This variable was not assessed in all participants. The denominator is listed next to the variable.

Table 2. Clinical presentation at admission of COVID-19 RT-PCR (+) and RT-PCR (−) groups.

Clinical Presentation at Admission (*n* = 2998)	COVID-19 PCR Positive *n* = 2292	COVID-19 PCR Negative *n* = 706	*p*-Value *,3
Days of symptoms **,4	7 (4–10) [2290]	8 (5–11) [704]	<0.001
Fever **	1286/2214 (58.1%)	374/664 (56.3%)	0.421
Cough **	1552/2210 (70.2%)	495/673 (73.6%)	0.096
Productive cough **	121/1310 (9.2%)	37/388 (9.5%)	0.859
Dyspnea **	1713/2219 (77.2%)	556/676 (82.2%)	0.005
Headache **	421/2160 (19.5%)	128/642 (19.9%)	0.802
Runny nose **	235/1924 (12.2%)	75/524 (14.3%)	0.200
Myalgias **	685/2176 (31.5%)	214/650 (32.9%)	0.488
Fatigue **	586/2000 (29.3%)	180/552 (32.6%)	0.133
Anosmia **	311/2165 (14.4%)	88/644 (13.7%)	0.655
Ageusia **	312/2158 (14.5%)	109/644 (16.9%)	0.124
Odynophagia **	147/2146 (6.8%)	52/642 (8.1%)	0.281
Diarrhea **	297/2164 (13.7%)	74/640 (11.6%)	0.156
Nausea **	242/2153 (11.2%)	80/641 (12.5%)	0.388
Vomit **	102/1460 (7.0%)	43/429 (10.0%)	0.038
Abdominal pain **	84/1907 (4.4%)	37/512 (7.2%)	0.009
Altered mental status **	130/1908 (6.8%)	41/514 (8.0%)	0.361

3.* *p* values were calculated using the nonparametric Mann-Whitney U command in Python (v 0.24.1), version 15.0, that tests for trend across ordered groups. Comparing COVID-19 RT-PCR positive and negative groups. 4.** This variable was not assessed in all participants. The denominator is listed next to the variable.

An analysis comparing the days of COVID-19 symptoms on hospital admission among different comorbidities was performed to investigate whether the different rates of comorbidities between COVID-19 RT-PCR(+) group and COVID-19 RT-PCR(−) group were due to patients with a certain comorbidity arriving earlier to the hospital, as the probability of a positive test increases with the decrease in days of symptoms on hospital admission [14]. Median days of COVID-19 symptoms on hospital admission differed in some subgroups, according to certain baseline diseases: Diabetes vs. No Diabetes (7 and 7, $p = 0.54$, 95% CI Figure 1), Cancer vs. No Cancer (4 and 8, $p < 0.001$, 95% CI Figure 1), Hypertension vs. No Hypertension (7 and 7, $p = 0.26$, 95% CI Figure 1), Asthma vs. No Asthma (7 and 7, $p = 0.46$, 95% CI Figure 1), Chronic Kidney Disease vs. No Chronic Kidney Disease (6 and 7, $p < 0.001$, 95% CI Figure 1) and Cardiovascular Disease vs. No Cardiovascular Disease (7 and 7, $p < 0.001$, 95% CI Figure 1).

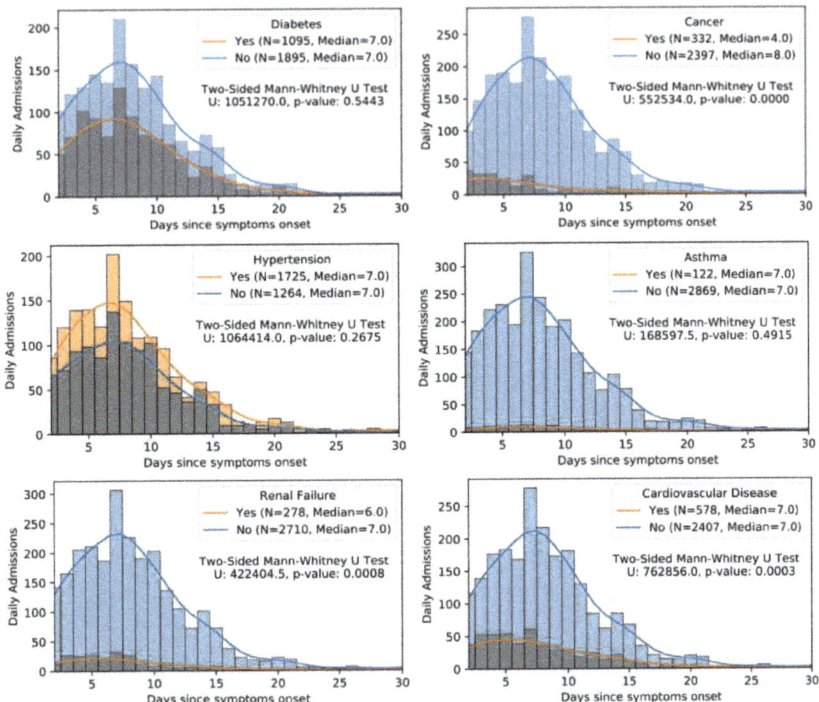

Figure 1. The median days of COVID-19 symptoms on hospital admission was: diabetes vs. no diabetes (7 and 7, $p = 0.54$), cancer vs. no cancer; (4 and 8, $p < 0.001$), hypertension vs. no hypertension (7 and 7, $p = 0.26$), asthma vs. no asthma (7 and 7, $p = 0.46$), chronic kidney disease vs. no chronic kidney disease (6 and 7, $p < 0.001$), and cardiovascular disease vs. no cardiovascular disease (7 and 7, $p < 0.001$).

Patients with negative RT-PCR presented more dyspnea (82.2% vs. 77.2%) and gastrointestinal symptoms than patients with positive RT-PCR (Table 2). All other symptoms were equal between the two groups. Physical examination between the two groups was similar, with a small but significant difference in oxygen saturation (Table 3). The median of oxygen saturation in the COVID-19 RT-PCR(+) group and COVID-19 RT-PCR(−) group was 94%, but the distribution of values was slightly different with lower oxygen saturation in the COVID-19 RT-PCR(-) group ($p = 0.046$, 95% CI).The SAPS score, which estimates the probability of mortality for ICU patients, was comparable in both groups.

Table 3. Physical examination at admission of COVID-19 RT-PCR (+) and RT-PCR (−) groups.

Physical Examination at Admission	COVID-19 PCR Positive n = 2292	COVID-19 PCR Negative n = 706	p-Value *[,5]
Heart rate **[,6] (beats/min)	88 (78–99) [2288]	88 (78–101) [706]	0.585
Respiratory rate ** (breaths/min)	24 (20–28) [2279]	23 (20–28) [704]	0.236
Temperature ** (°C)	36 (36–37) [2288]	36 (36–37) [703]	0.365°
Systolic blood pressure (mmHg) **	122 (110–140) [2286]	123 (110–140) [706]	0.688
Diastolic blood pressure (mmHg) **	76 (66–83) [2286]	76 (68–84) [706]	0.482
Oxygen saturation (%) **	94 (91–96) [2283]	94 (91–97) [706]	0.046
Simplified Acute Physiology Score III **	64 (53–77) [1444]	65 (53–76) [300]	0.631

[5,] * p values were calculated using the nonparametric Mann-Whitney U command in Python (v 0.24.1), version 15.0, that tests for trend across ordered groups. [6,] ** This variable was not assessed in all participants. The denominator is listed next to the variable.

3.2. *Laboratory Tests*

Patients in the RT-PCR(−) group had a higher number of circulating leukocytes, lymphocytes, and D-dimer, and lower CRP values (Table 4). The presence of antibodies against the SARS-CoV-2 was detectable in a higher percentage of patients in the RT-PCR(+) group compared to the RT-PCR(−) group (88.4% vs. 78.6%, $p < 0.001$, 95% CI, Table 4. Altogether, the data presented above suggest that both experimental groups are similar in disease severity; however, some differences in previous diseases' existence and laboratory tests were detected. As an attempt to establish baseline COVID-19 seropositivity in São Paulo, we checked the official data report released periodically by the mayor's office. It was found that the seroprevalence in this population was 11.4% as of June 2020 and 13.3% as of August 2020. Data related to previous months were unavailable in the official records of São Paulo state administration [15]. As the majority of our cohort was enrolled before June 2020, less than 11.4% of included patients were expected to have a baseline positive COVID-19 serology.

Table 4. Laboratory tests at admission of COVID-19 RT-PCR (+) and RT-PCR (−) groups.

Laboratory Tests at Admission	COVID-19 PCR Positive n = 2292	COVID-19 PCR Negative n = 706	p-Value *[,7]
Leukocytes **[,8] (thousand per mm^3)	8.22 (5.76–11.79) [2243]	9.19 (6.56–13.19) [690]	<0.001
Lymphocytes ** (thousand per mm^3)	0.86 [0.57–1.22] [2234]	1.03 [0.68–1.5] [689]	<0.001
C-reactive protein ** (mg/dL)	128 (64–225) [2090]	107 (50–201) [633]	<0.001
D-dimer ** (μg/mL)	1430 (766–4070) [1871]	1550 (875–5162) [585]	0.008
Positivity rate for COVID-19 serology **	344/389 (88.4%)	276/351 (78.6%)	<0.001

[7,] * p values were calculated using the nonparametric Mann-Whitney U command in Python (v 0.24.1), version 15.0, that tests for trend across ordered groups. [8,] ** This variable was not assessed in all participants. The denominator is listed next to the variable.

Patients in the RT-PCR(+) group were admitted to ICU at higher rates (67.1% vs. 50.0% ($p < 0.001$, 95% CI Table 1)), were submitted to endotracheal intubation more frequently (55.9% vs. 40.5% ($p < 0.001$, 95% CI Table 1)), and stayed longer in hospital (13 days vs. 9 days ($p < 0.001$, 95% CI Table 1)) when compared to patients in the RT-PCR(−) group (Table 5). Moreover, the in-hospital and 28-day mortality rates were higher in patients with positive RT-PCR tests ($p < 0.001$, 95% CI).

The COVID-19 RT-PCR(+) group, when compared to the RT-PCR(−) group had a hazard ratio (HR) of 1.44 (95% CI 1.26–1.65, $p < 0.005$) for intubation (Figure 2) whereas for 28-day in-hospital mortality the HR was 1.09 (95% confidence interval (CI) 0.91–1.31, $p = 0.33$) (Figure 3). In a multivariate model adjusted for age, sex, smoking status, hypertension, diabetes, asthma, and cancer, a positive RT-PCR was independently associated with increased risk of intubation (adjusted odds ratio (aOR), 2.04; 95% CI, 1.70–2.44; $p < 0.001$, Table 2) when compared to patients with a negative RT-PCR. Additionally, in a multivariate model adjusted for age, sex, body mass index, smoking status, hypertension, cardiovascular disease, chronic obstructive disease, chronic kidney disease, and cancer, the risk of a 28-day in-hospital mortal was higher in patients in the COVID-19 RT-PCR(+) group compared to the COVID-19 RT-PCR(−) group (aOR, 1.75; 95% CI, 1.41–2.16; $p < 0.001$, Table 6).

Table 5. Clinical outcomes of COVID-19 RT-PCR (+) and RT-PCR (−) groups [9].

Clinical Outcomes	COVID-19 PCR Positive $n = 2292$	COVID-19 PCR Negative $n = 706$	p-Value *,10
ICU care	1537/2292 (67.1%)	353/706 (50.0%)	<0.001
Intubation **,11	1207/2156 (55.9%)	250/617 (40.5%)	<0.001
Days until intubation **	8 (5–11) [1023]	8 (5–12) [189]	0.764
Mortality	842/2292 (36.7%)	161/706 (22.8%)	<0.001
28-Day mortality	726/2290 (31.7%)	146/706 (20.7%)	<0.001
Days until death **	21 (14–30) [839]	19 (11–28) [159]	0.051
Admission until death **	13 (8–21) [840]	10 (4–18) [159]	<0.001
Length of stay	13 (7–21) [2292]	9 (5–15) [706]	<0.001

[9] Variables are expressed as number (%) or median (interquartile range). Bolded values indicated variables with statistically significant associations (significance level of 0.05). [10] * p values were calculated using the nonparametric Mann-Whitney U command in Python (v 0.24.1), version 15.0, that tests for trend across ordered groups. [11] ** This variable was not assessed in all participants. The denominator is listed next to the variable.

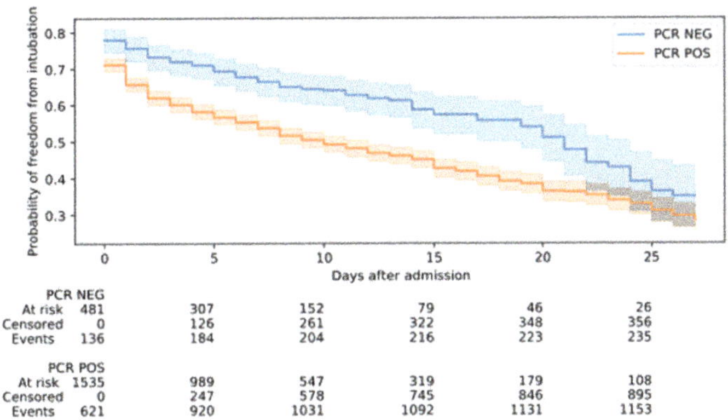

Figure 2. A positive RT-PCR test result was associated with a hazard ratio (HR) of 1.44 (95% confidence interval (CI) 1.26–1.65, p value < 0.005) for intubation.

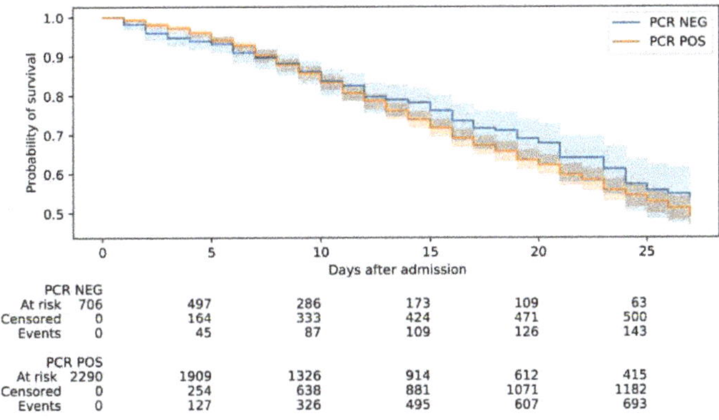

Figure 3. A positive RT-PCR was associated with a hazard ratio (HR) of 1.09 (95% confidence interval (CI) 0.91–1.31, $p = 0.33$) for in-hospital mortality.

Table 6. Multivariate Logistic Regression Models of Factors Associated with Intubation and 28-day In-hospital Mortality [12].

	Intubation		28-Day Mortality	
	Adjusted OR (95% CI)	*p*-Value *,[13]	Adjusted OR (95% CI)	*p*-Value *
Positive SARS-CoV-2 RT-PCR	2.04 (1.70–2.44)	<0.001	1.75 (1.41–2.16)	<0.001
Age, years	1.01 (1.00–1.01)	0.04	1.05 (1.04–1.05)	<0.001
Male Sex	1.38 (1.19–1.61)	<0.001	1.37 (1.15–1.63)	<0.001
Current smoker **,[14]	1.93 (1.41–2.65)	<0.001		
Comorbidities				
Hypertension **	1.22 (1.03–1.45)	0.02		
Cardiovascular disease **			1.22 (1.01–1.46)	0.03
Diabetes **	1.35 (1.14–1.59)	<0.001		
Asthma **	0.63 (0.43–0.94)	0.02		
Active cancer **	0.61 (0.48–0.77)	<0.001	1.8 (1.40–2.31)	<0.001

[12] Only variables that had a significant association with intubation or mortality in a univariate logistic regression model were included in the corresponding multivariate model. Empty cells and absence in the table indicate that the variable was not associated with the corresponding outcome in the univariate logistic regression model. Abbreviations: CI, confidence interval; OR, odds ratio. [13] * *p* values were calculated using the nonparametric Mann-Whitney U command in Python (v 0.24.1), version 15.0, that tests for trend across ordered groups. [14] ** This variable was not assessed in all participants. The denominator is listed next to the variable.

An analysis of a subgroup of patients without comorbidities that differed between the COVID-19 RT-PCR(+) and COVID-19 RT-PCR(−) was also performed. Accordingly, only patients without cardiovascular disease, diabetes, chronic kidney disease, peripheral vascular disease, cancer, and hematologic malignancy were enrolled. In this subgroup, the positive RT-PCR association with intubation (64.4% vs. 47.6%, $p < 0.001$, Table 7) and 28-day mortality (30.8% vs. 19.3%, $p = 0.002$, Table 7) was maintained.

Table 7. Comparison of demographic characteristics, frequency of comorbidities, and outcomes in the COVID-19 RT-PCR (+) and COVID-19 RT-PCR (−) subgroup of patients without cardiovascular disease, diabetes, chronic kidney disease, peripheral vascular disease, cancer, and hematologic malignancy [15].

Variable	COVID-19 PCR Positive $n = 595$	COVID-19 PCR Negative $n = 207$	*p*-Value *,[16]
Age, years	62.2 (54.4–70.9) [595]	61.3 (53.3–70.7) [207]	0.451
Male Sex	356/595 (59.8%)	119/207 (57.5%)	0.555
Body mass index **	26.6 (23.3–31.2) [478]	26.0 (23.5–30.5) [153]	0.825
Former or current smoker **	123/594 (20.7%)	43/207 (20.8%)	0.984
Current smoker	55/595 (9.2%)	12/207 (5.8%)	0.123
Previous Diseases			
Hypertension	267/595 (44.9%)	90/207 (43.5%)	0.728
Cerebrovascular disease	23/595 (3.9%)	11/207 (5.3%)	0.374
Arrhythmia **	14/594 (2.4%)	6/206 (2.9%)	0.660
Chronic obstructive pulmonary disease **,[17]	42/594 (7.1%)	11/207 (5.3%)	0.382
Asthma	17/595 (2.9%)	7/207 (3.4%)	0.703
End-stage renal disease	11/595 (1.8%)	3/207 (1.4%)	0.706
Rheumatologic disease	15/595 (2.5%)	7/207 (3.4%)	0.514
Liver disease	14/595 (2.4%)	3/207 (1.4%)	0.438
Human immunodeficiency virus infection	6/595 (1.0%)	2/207 (1.0%)	0.959

Table 7. *Cont.*

Variable	COVID-19 PCR Positive n = 595	COVID-19 PCR Negative n = 207	p-Value *,16
Outcomes			
ICU care	421/595 (70.8%)	112/207 (54.1%)	<0.001
Intubation **	358/556 (64.4%)	81/170 (47.6%)	<0.001
Days until intubation **	8.0 (5.0–12.0) [275]	9.0 (7.0–12.0) [50]	0.153
Mortality	218/595 (36.6%)	48/207 (23.2%)	<0.001
28-Day mortality	183/595 (30.8%)	40/207 (19.3%)	0.002
Days until death **	23.0 (17.0–30.0) [218]	23.0 (15.5–32.8) [46]	0.845
Admission until death **	15.0 (10.0–22.8) [218]	12.0 (6.2–22.5) [46]	0.190
Length of stay	14.0 (7.0–23.0) [595]	10.0 (5.5–19.0) [207]	<0.001

[15] Variables are expressed as number (%) or median (interquartile range). Bolded values indicated variables with statistically significant associations (significance level of 0.05). [16] * p values were calculated using the nonparametric Mann-Whitney U command in Python (v 0.24.1), version 15.0, that tests for trend across ordered groups. [17] ** This variable was not assessed in all participants. The denominator is listed next to the variable.

Importantly, only missing data completely at random were observed. Missing data at random and not at random were not observed. A strict evaluation of the dataset highlighted that missing data occurred due to insufficient filling of electronic medical records in punctual episodes. Accordingly, complete case analysis was used to minimize the biases derived from missing data.

4. Discussion

In this study, COVID-19 patients with at least one positive SARS-CoV-2 RT-PCR result were more likely to be intubated or die during hospitalization when compared to patients that had only clinical and radiological criteria for COVID-19 diagnosis. This association persisted even after an adjustment for age and comorbidities was performed.

The first question to be raised is whether participants in the RT-PCR(−) group really had COVID-19. This issue was approached by checking the antibody production in these patients. Although we did not have data from all patients, some patients (n = 351 patients) were tested and 78.6% presented serologic conversion after hospital admission. Even though this rate is lower than the one found in the RT-PCR(+) group, it is significant. This is indirect evidence that the clinical and radiological presumptive diagnosis was probably accurate in most patients in the COVID-19 RT-PCR(−) group. These data are in accordance with other authors showing that CT scan is more sensitive than RT-PCR for COVID-19 diagnosis [16] as RT-PCR test accuracy may be questioned. A total of five studies, involving 957 patients, demonstrated that false negative results range from 2 to 29% (equating to sensitivity of 71–98%) [7]. Although the authors judged their evidence as low, due to the risk of bias, indirectness, and inconsistency issues, the occurrence of false negative SARS-CoV-2 RT-PCR results must be considered.

There are several reasons that can underlie false-negative RT-PCR nasopharyngeal SARS-CoV-2 results, including preanalytical steps (conservation of samples, time until being sent to the laboratory, and training of personnel) [17], the number of additional RT-PCR assays performed [7], and insufficient viral load (possibly influenced by time from the onset of symptoms) [13]. In our institution, the protocol used to test for SARS-CoV-2 infection requires at least 40 viral RNA copies per milliliter to detect viral presence in 95% of cases (defined as limit of detection (LOD)) suggesting that viral load may be one of the factors contributing to the overall number of false negative tests in this cohort.

Likewise, another interesting aspect of this study is the relationship between the high number of participants in the RT-PCR(−) group and the RT-PCR protocol used. Corman et al. [13] suggested that the first-line protocol should include an RNA analysis of the viral gene E followed by an analysis of the viral gene RdRp. Alternatively, viral gene N could also be used instead of gene RdRp, although it had a higher LOD. In the present study, as it may occur in several other health care institutions throughout other BRICS' nations, limitation of resources and logistics prevented the laboratory facility from using gene RdRp in the RT-PCR protocol. Due to the difference between the LODs, one may hypothesize that the protocol of using gene N contributed to the occurrence of false negatives, increasing the number of participants in the RT-PCR(−) group. However, as mentioned above, we believe that this potential limitation may be present in other institutions, which grants relevance to the presenting results as it depicts real, daily-care difficulties, and may guide practitioners during clinical decision-making in the context of limited hospital and laboratorial resources (e.g., ICU beds, mechanical ventilation, and updated diagnostic protocols, etc.).

Patients with cardiovascular disease, peripheral vascular disease, diabetes, chronic kidney disease, active cancer, and hematologic malignancy were more frequent in the COVID-19 RT-PCR(+) group. A possible explanation for the higher rate of chronic comorbidities in the COVID-19 RT-PCR(+) group is that those patients had a more severe spectrum of COVID-19 [18] and presented earlier to the emergency department, which could have led to a higher chance of having a positive SARS-CoV-2 RT-PCR test, as the probability of a positive test increases with the decrease in days of symptoms on hospital admission [13]. This may be a feasible explanation for the higher cancer rate in the COVID-19 RT-PCR(+) group (13.1% vs. 8.8%, $p = 0.004$, Table 1) as the median days of symptoms on hospital admission for cancer and non-cancer patients are 4 and 8, respectively (Figure 1). However, this significant difference of median days of symptoms on hospital admission is not seen among other comorbidities (Figure 1). Additionally, the higher rate of comorbidities in the COVID-19 RT-PCR(+) group could explain the higher risk of intubation and mortality in that group, but the overall association between a positive SARS-CoV-2 RT-PCR and 28-day mortality and intubation persisted even after an adjustment for comorbidities, suggesting an independent association between SARS-CoV-2 RT-PCR and COVID-19 outcomes (Table 2). We are aware that unmeasured imbalances can persist after a statistical adjustment, especially in observational studies.

Another hypothetical explanation for the higher rate of some comorbidities in the COVID-19 RT-PCR(+) group is a systemic difficulty to clear the SARS-CoV-2 viral load. Our results are in line with previous evidence, which demonstrated that COVID-19 patients with cardiovascular disease, chronic kidney disease, active cancer, and hematologic malignancy present a higher viral load on hospital admission and, thus, a higher probability of a positive RT-PCR test [19]. Reasons for higher viral loads specifically in these populations are not fully understood and warrant further investigation. However, in previous studies, chronic obstructive pulmonary disease and age were also associated with higher viral load [19] whereas in the present study they were not different among COVID-19 RT-PCR(+) and COVID-19 RT-PCR(−) groups. This may be caused by the limitations of evaluating only SARS-CoV-2 RT-PCR results.

Although the COVID-19 RT-PCR(+) group had a higher rate of comorbidities, after an adjustment for the factors that are associated with a worse COVID-19 outcome, the aforementioned group maintained an excessive incidence of intubation and mortality when compared to the COVID-19 RT-PCR(−) group. Similarly, higher SARS-CoV-2 viral loads are independently associated with intubation and mortality [19]. A higher viral load in the COVID-19 RT-PCR(+) group could be a possible explanation for the superior intubation and mortality rates in that group.

Other limitations must also be considered. The number of RT-PCR assays performed per patient, a procedure that is known to increase test sensitivity [7], was not reported due to a failure in our database. Nonetheless, according to our institutional protocol, patients with suspected COVID-19 infection but a negative initial RT-PCR result were submitted to at least two other consecutive RT-PCR tests. Therefore, all patients in the RT-PCR(−) group had a minimum of two negative tests.

Radiological chest CT scan findings were not provided due to a limitation of our database design. Another limitation is the imprecise definition of the criteria used to delimitate the RT-PCR(−) group. Our task force of researchers was instructed to input into our dataset if at discharge the attending physicians considered that the patient had a clinical and radiological presumptive diagnosis of COVID-19, even with a negative SARS-CoV-2 RT-PCR result, after ruling out other differential diagnoses. Still, as the reference public tertiary center in Brazil, we believe that this is reliable information. Our doctors are highly trained to identify suspicious clinical and radiological cases of COVID-19 and to rule-out differential diagnoses. Another limitation is that patients were enrolled from March to August 2020. As of August 2020, in Brazil, mainly the following COVID-19 variants were present: B.1.1.33 (37.8%), B.1.1.28 (32.5%), and B.1.1 (16.4%). The primers used to detect SARS-CoV-2 in our hospital were validated by our Laboratory Division (accredited by the College of American Pathologists) and in line with the circulating SARS-CoV-2 strains in our country. Importantly, the gamma variant is not present in this study as it was only reported in Sao Paulo in January 2021 [20].We do not know whether these findings are applicable for any new variants including alpha, beta, delta, gamma, and omicron.

This study has some strengths too. Many patients were enrolled (n = 2998), decreasing the risk of a type II error (stating that there is not an effect or difference when one exists). In terms of data collection, although there were some missing values, only data missing completely at random were observed. Importantly, the results reported in this study with hospitalized and symptomatic patients are not reproducible in an outpatient setting. In conclusion, we found that the presence of at least one positive SARS-CoV-2 RT-PCR result was independently associated with intubation or 28 days death during hospitalization when compared to patients who had only a presumptive clinical and radiological criterion for COVID-19 diagnosis without a second, more probable differential diagnosis. Of note, CT findings are sensitive but not specific for COVID-19 [16], as a result, only patients with both clinical suspicion for COVID-19 and a characteristic CT scan were enrolled in the COVID-19 RT-PCR(−) group. These findings suggest that patients clinically considered to have COVID-19 but with a negative SARS-CoV-2 RT-PCR result may have less risk of COVID-19-related unfavorable outcomes.

Author Contributions: Conceptualization, M.C.S.M., J.C.F., C.R.R.d.C. and H.P.S.; methodology, H.P.S., M.C.S.M., D.D.M., D.V.S.P. and C.R.R.d.C.; formal analysis, H.P.S., M.C.S.M., D.D.M. and D.V.S.P.; investigation, M.C.S.M., H.P.S., D.V.S.P., L.M.G.G., J.F.M. and J.C.A.; resources, C.R.R.d.C., J.C.F., H.P.S. and Emergency USP COVID-19 Group; data curation, M.C.S.M., D.V.S.P., D.D.M., H.P.S. and HCFMUSP COVID-19 Study Group; writing—original draft preparation, M.C.S.M., D.V.S.P., H.P.S. and D.D.M.; writing—review and editing, C.R.R.d.C., J.F.M., J.C.F., J.C.A. and H.P.S.; visualization, C.R.R.d.C., J.F.M., J.C.F., J.C.A. and H.P.S.; supervision, M.C.S.M., D.V.S.P. and H.P.S.; project administration, M.C.F., I.O.M., K.R.d.S. and V.C.J.; funding acquisition, H.P.S. and C.R.R.d.C. All authors have read and agreed to the published version of the manuscript.

Funding: This research was financially supported by Fundação de Amparo à Pesquisa do Estado de São Paulo, grants number #2020/04.738-8 and #2016/14.566-4.

Institutional Review Board Statement: The study was conducted according to the guidelines of the Declaration of Helsinki and approved by the Institutional Review Board and the Ethics Committee of the Hospital das Clínicas da Universidade de São Paulo (HC-FMUSP/protocol number: 30417520.0.0000.0068).

Informed Consent Statement: The institutional Ethics Committee approved the study with a waiver of informed consent under protocol number CAAE: 30417520.0.0000.0068. The informed consent was waived because the signature on the informed consent document would be the only record linking the subject to the research and the principal risk of harm to the subject would be a breach of confidentiality.

Data Availability Statement: Due to its sensitive nature involving patient health information, the data presented in this study are available upon a reasonable request.

Conflicts of Interest: The authors declare no conflict of interest.

References

1. Brandão Neto, R.A.; Marchini, J.F.; Marino, L.O.; Alencar, J.C.; Lazar Neto, F.; Ribeiro, S.; Salvetti, F.V.; Rahhal, H.; Gomez Gomez, L.M.; Bueno, C.G.; et al. Mortality and other outcomes of patients with coronavirus disease pneumonia admitted to the emergency department: A prospective observational Brazilian study. *PLoS ONE* **2021**, *16*, e0244532.
2. Goyal, P.; Choi, J.J.; Pinheiro, L.C.; Schenck, E.J.; Chen, R.; Jabri, A.; Satlin, M.J.; Campion Jr, T.R.; Nahid, M.; Ringel, J.B.; et al. Clinical Characteristics of COVID-19 in New York City. *N. Engl. J. Med.* **2020**, *382*, 2372–2374. [CrossRef] [PubMed]
3. Wu, C.; Chen, X.; Cai, Y.; Zhou, X.; Xu, S.; Huang, H.; Zhang, L.; Zhou, X.; Du, C.; Zhang, Y.; et al. Risk Factors Associated With Acute Respiratory Distress Syndrome and Death in Patients With Coronavirus Disease 2019 Pneumonia in Wuhan, China. *JAMA Intern. Med.* **2020**, *180*, 934–943. [CrossRef] [PubMed]
4. World Health Organization (WHO). *Clinical Management of COVID-19—Interim Guidance*; World Health Organization: Geneva, Switzerland, 27 May 2020. Available online: https://apps.who.int/iris/handle/10665/332196 (accessed on 6 January 2022).
5. Woloshin, S.; Patel, N.; Kesselheim, A.S. False Negative Tests for SARS-CoV-2 Infection—Challenges and Implications. *N. Engl. J. Med.* **2020**, *383*, e38. [CrossRef]
6. Floriano, I.; Silvinato, A.; Bernardo, W.M.; Reis, J.C.; Soledade, G. Accuracy of the Polymerase Chain Reaction (PCR) test in the diagnosis of acute respiratory syndrome due to coronavirus: A systematic review and meta-analysis. *Rev. Assoc. Médica Bras.* **2020**, *66*, 880–888. [CrossRef]
7. Arevalo-Rodriguez, I.; Buitrago-Garcia, D.; Simancas-Racines, D.; Zambrano-Achig, P.; Del Campo, R.; Ciapponi, A.; Sued, O.; Martinez-Garcia, L.; Rutjes, A.W.; Low, N.; et al. False-negative results of initial RT-PCR assays for COVID-19: A systematic review. *PLoS ONE* **2020**, *15*, e0242958. [CrossRef] [PubMed]
8. Baicry, F.; Le Borgne, P.; Fabacher, T.; Behr, M.; Lemaitre, E.L.; Gayol, P.A.; Harscoat, S.; Issur, N.; Garnier-Kepka, S.; Ohana, M.; et al. Patients with Initial Negative RT-PCR and Typical Imaging of COVID-19: Clinical Implications. *J. Clin. Med.* **2020**, *9*, 3014. [CrossRef] [PubMed]
9. Fistera, D.; Härtl, A.; Pabst, D.; Manegold, R.; Holzner, C.; Taube, C.; Dolff, S.; Schaarschmidt, B.M.; Umutlu, L.; Kill, C.; et al. What about the others: Differential diagnosis of COVID-19 in a German emergency department. *BMC Infect. Dis.* **2021**, *21*, 969. [CrossRef] [PubMed]
10. Simpson, S.; Kay, F.U.; Abbara, S.; Bhalla, S.; Chung, J.H.; Chung, M.; Henry, T.S.; Kanne, J.P.; Kligerman, S.; Ko, J.P.; et al. Radiological Society of North America Expert Consensus Statement on Reporting Chest CT Findings Related to COVID-19: Endorsed by the Society of Thoracic Radiology, the American College of Radiology, and RSNA—Secondary Publication. *J. Thorac. Imaging* **2020**, *35*, 219–227. [CrossRef] [PubMed]
11. Ai, T.; Yang, Z.; Hou, H.; Zhan, C.; Chen, C.; Lv, W.; Tao, Q.; Sun, Z.; Xia, L. Correlation of Chest CT and RT-PCR Testing for Coronavirus Disease 2019 (COVID-19) in China: A Report of 1014 Cases. *Radiology* **2020**, *296*, E32–E40. [CrossRef] [PubMed]
12. Harris, P.A.; Taylor, R.; Thielke, R.; Payne, J.; Gonzales, N.; Conde, J.G. Research electronic data capture (REDCap)—A metadata-driven methodology and workflow process for providing translational research informatics support. *J. Biomed. Inform.* **2009**, *42*, 377–381. [CrossRef]
13. Corman, V.M.; Landt, O.; Kaiser, M.; Molenkamp, R.; Meijer, A.; Chu, D.K.; Bleicker, T.; Brünink, S.; Schneider, J.; Schmidt, M.L.; et al. Detection of 2019 novel coronavirus (2019-nCoV) by real-time RT-PCR. *Eurosurveillance* **2020**, *25*, 2000045. [CrossRef] [PubMed]
14. Wikramaratna, P.S.; Paton, R.S.; Ghafari, M.; Lourenço, J. Estimating the false-negative test probability of SARS-CoV-2 by RT-PCR. *Eurosurveillance* **2020**, *25*, 2000568. [CrossRef] [PubMed]
15. Coronavírus—Dados Completos. Available online: https://www.seade.gov.br/coronavirus/ (accessed on 6 January 2022).
16. Mair, M.D.; Hussain, M.; Siddiqui, S.; Das, S.; Baker, A.; Conboy, P.; Valsamakis, T.; Uddin, J.; Rea, P. A systematic review and meta-analysis comparing the diagnostic accuracy of initial RT-PCR and CT scan in suspected COVID-19 patients. *Br. J. Radiol.* **2021**, *94*, 20201039. [CrossRef] [PubMed]
17. Payne, D.; Newton, D.; Evans, P.; Osman, H.; Baretto, R. Preanalytical issues affecting the diagnosis of COVID-19. *J. Clin. Pathol.* **2021**, *74*, 207–208. [CrossRef] [PubMed]
18. Argenziano, M.G.; Bruce, S.L.; Slater, C.L.; Tiao, J.R.; Baldwin, M.R.; Barr, R.G.; Chang, B.P.; Chau, K.H.; Choi, J.J.; Gavin, N. Characterization and clinical course of 1000 patients with coronavirus disease 2019 in New York: Retrospective case series. *BMJ* **2020**, *369*, m1996. [CrossRef] [PubMed]
19. Magleby, R.; Westblade, L.F.; Trzebucki, A.; Simon, M.S.; Rajan, M.; Park, J.; Goyal, P.; Safford, M.M.; Satlin, M.J. Impact of Severe Acute Respiratory Syndrome Coronavirus 2 Viral Load on Risk of Intubation and Mortality Among Hospitalized Patients With Coronavirus Disease 2019. *Clin. Infect. Dis.* **2020**, *73*, e4197–e4205. [CrossRef] [PubMed]
20. Dashboard—Genomahcov—Fiocruz. Available online: http://www.genomahcov.fiocruz.br/frequencia-das-principais-linhagens-do-SARS-CoV-por-mes-de-amostragem/ (accessed on 6 January 2022).

Article

Clinical Characterization and Genomic Analysis of Samples from COVID-19 Breakthrough Infections during the Second Wave among the Various States of India

Nivedita Gupta [1], Harmanmeet Kaur [1], Pragya Dhruv Yadav [2,*], Labanya Mukhopadhyay [1], Rima R. Sahay [2], Abhinendra Kumar [2], Dimpal A. Nyayanit [2], Anita M. Shete [2], Savita Patil [2], Triparna Majumdar [2], Salaj Rana [1], Swati Gupta [1], Jitendra Narayan [1], Neetu Vijay [1], Pradip Barde [3], Gita Nataraj [4], Amrutha Kumari B. [5], Manasa P. Kumari [5], Debasis Biswas [6], Jyoti Iravane [7], Sharmila Raut [8], Shanta Dutta [9], Sulochana Devi [10], Purnima Barua [11], Piyali Gupta [12], Biswa Borkakoty [13], Deepjyoti Kalita [14], Kanwardeep Dhingra [15], Bashir Fomda [16], Yash Joshi [2], Kapil Goyal [17], Reena John [18], Ashok Munivenkatappa [19], Rahul Dhodapkar [20], Priyanka Pandit [2], Sarada Devi [21], Manisha Dudhmal [2], Deepa Kinariwala [22], Neeta Khandelwal [23], Yogendra Kumar Tiwari [24], Prabhat Kiran Khatri [25], Anjli Gupta [26], Himanshu Khatri [27], Bharti Malhotra [28], Mythily Nagasundaram [29], Lalit Dar [30], Nazira Sheikh [31], Jayanthi Shastri [32], Neeraj Aggarwal [1] and Priya Abraham [2]

1 Indian Council of Medical Research, V. Ramalingaswami Bhawan, Ansari Nagar, New Delhi 110029, India; drguptanivedita@gmail.com (N.G.); harmanmeet.kaur@gmail.com (H.K.); labanya.mukhopadhyay@gmail.com (L.M.); salajrana05@gmail.com (S.R.); 07guptaswati@gmail.com (S.G.); jitunarayan@gmail.com (J.N.); drneetuvijay@gmail.com (N.V.); aggarwal.n@icmr.gov.in (N.A.)

2 Indian Council of Medical Research-National Institute of Virology, Pune 411021, India; dr.rima.sahay@gmail.com (R.R.S.); abhinendra.biotech@gmail.com (A.K.); nyayanit.dimpal@gmail.com (D.A.N.); anitaaich2008@gmail.com (A.M.S.); varshapatil111@yahoo.com (S.P.); triparna.majumdar@gmail.com (T.M.); yashjos1401@gmail.com (Y.J.); priyanka.pb83@gmail.com (P.P.); dudhmalmanisha23@gmail.com (M.D.); priya.abraham@icmr.gov.in (P.A.)

3 Viral Research and Diagnostic Laboratory, National Institute of Research in Tribal Health (NIRTH), Jabalpur 482003, India; pradip_barde@hotmail.com

4 Viral Research and Diagnostic Laboratory, Department of Microbiology, KEM Medical College, Mumbai 400012, India; gitanataraj@gmail.com

5 Viral Research and Diagnostic Laboratory, Department of Microbiology, Mysore Medical College, Mysore 570015, India; amrutakb@yahoo.co.in (A.K.B.); manasavinay22@gmail.com (M.P.K.)

6 Viral Research and Diagnostic Laboratory, Department of Microbiology, All India Institute of Medical Sciences, Bhopal 462020, India; debasis.microbiology@aiimsbhopal.edu.in

7 Viral Research and Diagnostic Laboratory, Government Medical College, Aurangabad 431001, India; jairavane@hotmail.com

8 Viral Research and Diagnostic Laboratory, Indira Gandhi Government Medical College, Nagpur 440012, India; sharmilakuber@gmail.com

9 Viral Research and Diagnostic Laboratory, National Institute of Cholera and Enteric Diseases, Kolkata 700010, India; drshantadutta@gmail.com

10 Viral Research and Diagnostic Laboratory, Regional Institute of Medical Sciences, Imphal 795004, India; sulo_khu@rediffmail.com

11 Viral Research and Diagnostic Laboratory, Jorhat Medical College, Jorhat 785001, India; drpurnimabarua@gmail.com

12 Viral Research and Diagnostic Laboratory, Mahatma Gandhi Memorial Medical College, Jamshedpur 831020, India; mgmvrdl@gmail.com

13 Viral Research and Diagnostic Laboratory, ICMR-Regional Medical Research Centre, Dibrugarh 786001, India; biswaborkakoty@gmail.com

14 Viral Research and Diagnostic Laboratory, All India Institutes of Medical Sciences, Rishikesh 249203, India; deep.micro@aiimsrishikesh.edu.in

15 Viral Research and Diagnostic Laboratory, Government Medical College, Amritsar 143001, India; kdmicrogmcasr@gmail.com

16 Viral Research and Diagnostic Laboratory, Sher-i-Kashmir Institute of Medical Sciences, Srinagar 190011, India; bashirfomda@gmail.com

17 Department of Virology, Postgraduate Institute of Medical Education and Research, Chandigarh 160012, India; kapilgoyalpgi@gmail.com

18 Viral Research and Diagnostic Laboratory, Government Medical College, Thrissur 680596, India; rejovi3@gmail.com

Citation: Gupta, N.; Kaur, H.; Yadav, P.D.; Mukhopadhyay, L.; Sahay, R.R.; Kumar, A.; Nyayanit, D.A.; Shete, A.M.; Patil, S.; Majumdar, T.; et al. Clinical Characterization and Genomic Analysis of Samples from COVID-19 Breakthrough Infections during the Second Wave among the Various States of India. *Viruses* **2021**, *13*, 1782. https://doi.org/10.3390/v13091782

Academic Editors: Burtram C. Fielding and Georgia Schäfer

Received: 9 July 2021
Accepted: 29 August 2021
Published: 7 September 2021

[19] ICMR-National Institute of Virology Field Unit, Bangalore 560011, India; ashokmphdns@gmail.com
[20] Viral Research and Diagnostic Laboratory, Jawaharlal Institute of Postgraduate Medical Education & Research, Puducherry 605006, India; rahuldhodapkar@gmail.com
[21] Viral Research and Diagnostic Laboratory, Government Medical College, Thiruvanthapuram 695011, India; sdevikl23@gmail.com
[22] Viral Research and Diagnostic Laboratory, B. J. Medical College, Ahmedabad 380016, India; poliobjmedical@gmail.com
[23] Viral Research and Diagnostic Laboratory, Government Medical College, Surat 395001, India; neetashokk@gmail.com
[24] Viral Research and Diagnostic Laboratory, Jhalawar Medical College, Jhalawar 326001, India; yogendratiwari2012@gmail.com
[25] Viral Research and Diagnostic Laboratory, Dr. Sampurnanand Medical College, Jodhpur 342003, India; drpkkhatri@yahoo.co.in
[26] Viral Research and Diagnostic Laboratory, Sarder Patel Medical College, Bikaner 334001, India; vrdlbikaner@gmail.com
[27] Viral Research and Diagnostic Laboratory, Department of Microbiology, GMERS Medical College, Himmatnagar 383001, India; microgmershmt@gmail.com
[28] Viral Research and Diagnostic Laboratory, Sawai Man Singh Medical College, Jaipur 302004, India; drbhartimalhotra@gmail.com
[29] Viral Research and Diagnostic Laboratory, Coimbatore Medical College, Coimbatore 641018, India; mythilynmicro@gmail.com
[30] Viral Research and Diagnostic Laboratory, All India Institute of Medical Sciences, Delhi 110029, India; lalitdaraiims@gmail.com
[31] Viral Research and Diagnostic Laboratory, Dr. V.M Government Medical College, Solapur 413003, India; vrdlsolapur@gmail.com
[32] Viral Research and Diagnostic Laboratory, Kasturba Hospital for Infectious Diseases, Mumbai 400011, India; jsshastri@gmail.com
* Correspondence: hellopragya22@gmail.com; Tel.: +91-20-2600-6111; Fax: +91-20-2612-2669

Abstract: From March to June 2021, India experienced a deadly second wave of COVID-19, with an increased number of post-vaccination breakthrough infections reported across the country. To understand the possible reason for these breakthroughs, we collected 677 clinical samples (throat swab/nasal swabs) of individuals from 17 states/Union Territories of the country who had received two doses (n = 592) and one dose (n = 85) of vaccines and tested positive for COVID-19. These cases were telephonically interviewed and clinical data were analyzed. A total of 511 SARS-CoV-2 genomes were recovered with genome coverage of higher than 98% from both groups. Analysis of both groups determined that 86.69% (n = 443) of them belonged to the Delta variant, along with Alpha, Kappa, Delta AY.1, and Delta AY.2. The Delta variant clustered into four distinct sub-lineages. Sub-lineage I had mutations in ORF1ab A1306S, P2046L, P2287S, V2930L, T3255I, T3446A, G5063S, P5401L, and A6319V, and in N G215C; Sub-lineage II had mutations in ORF1ab P309L, A3209V, V3718A, G5063S, P5401L, and ORF7a L116F; Sub-lineage III had mutations in ORF1ab A3209V, V3718A, T3750I, G5063S, and P5401L and in spike A222V; Sub-lineage IV had mutations in ORF1ab P309L, D2980N, and F3138S and spike K77T. This study indicates that majority of the breakthrough COVID-19 clinical cases were infected with the Delta variant, and only 9.8% cases required hospitalization, while fatality was observed in only 0.4% cases. This clearly suggests that the vaccination does provide reduction in hospital admission and mortality.

Keywords: breakthrough; COVID-19; VRDL; Delta and Delta plus variant; India; vaccine

1. Introduction

Severe acute respiratory syndrome coronavirus-2 (SARS-CoV-2) was reported from Wuhan, China, in December 2019 and rapidly spread across the globe. The World Health Organization (WHO) declared the disease caused by it, Coronavirus disease of 2019 (COVID-19) as a Public Health Emergency of International Concern on 11t March 2020. Since then, the virus has been continuously evolving, and the first major mutation was seen in its spike protein (D614G), which led to increased infectivity [1]. However, several new SARS-CoV-2 variants of concern (VOCs), i.e., Alpha (B.1.1.7), Beta (B.1.351), and Gamma (B.1.1.28.1), have been detected from United Kingdom, South Africa, and Brazil, respectively, from September to December 2020 and have also been reported from India [2–4]. Our earlier study of genomic surveillance from January to August 2020 showed the absence of VOC/variants under investigation (VUIs) and the presence of the G, GR, and GH clade in the country, with a number of mutations [5]. The global circulation of variants amplified the COVID-19 pandemic with increased transmissibility, enhanced severity of illness, diminished protection relative to previous SARS-CoV-2 variant infection, and lower response to vaccines and monoclonal antibodies [6–8].

Since the worldwide alert of VOCs, international travelers arriving at Indian airports from December 2020 to date were tracked and subjected to diagnostic testing by SARS-CoV-2 specific real-time reverse transcription-polymerase chain reaction (rRTPCR). Genomic surveillance led to the detection of VOCs, i.e., Alpha and Beta; variants of interest (VOIs), i.e., Eta (B.1.525), Kappa (B.1.617.1), and Zeta (B.1.1.28.2); and variant under monitoring, i.e., B.1.617.3 [2–5,9,10]. The recent emergence of the B.1.617 lineage has created a grave public health problem in India. The lineage evolved further to generate sub-lineages B.1.617.1, B.1.617.2, and B.1.617.3 [11]. The sub-lineage B.1.617.2 has gradually dominated the other variants, including B.1.617.1, B.617.3, and Alpha VOC in Maharashtra state [9,10]. This variant has further evolved into two new strains called Delta AY.1 and Delta AY.2. The AY.1 and AY.2 variants have been aggregated with Delta variant B.1.617.2 [12].

Several candidate vaccines have been developed using various platforms on fast-track mode. Many of them have been used in different countries across the globe under emergency use authorization (EUA). On 1 January and 2 January 2021, the National Regulatory Authority of India accorded restricted emergency use authorization to the viral vector vaccine developed by Oxford–AstraZeneca (Covishield, manufactured in India) and inactivated vaccine BBV152 (Covaxin), respectively. Subsequently Sputnik V received EUA on 13 April 2021. In India, the national COVID-19 vaccination program was launched on 16 January 2021. Up to 3 June 2021, 132,847,680 individuals had received one dose, while 45,623,351 individuals had received two doses [13]. The timeline for COVID-19 vaccination in India is graphically represented in Figure 1.

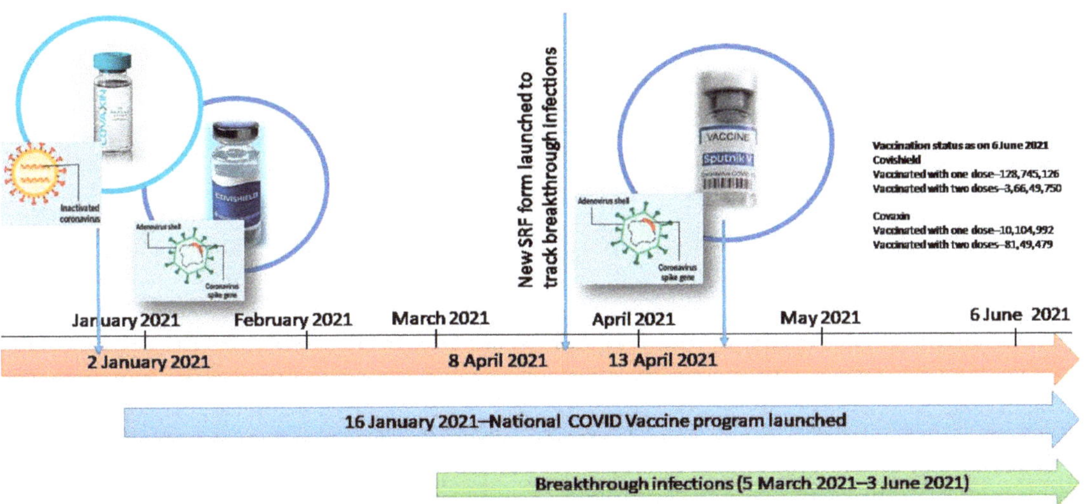

Figure 1. Graphical representation of timelines of vaccine emergency use authorization in India.

Protection offered by vaccines is being questioned following the emergence of VOCs and reduced real-world effectiveness of certain candidate vaccines against these variants.

Israel reported breakthrough COVID-19 infections in individuals immunized with the Pfizer vaccine on 9 April 2021 [14]. In India, we noted a second surge in the number of COVID-19 cases from March 2021, and this was followed by a devastating second wave. Further in April 2021, reduction in neutralization capacity of Covishield/AstraZeneca vaccinated sera against the B.1.1.7 variant as compared with the ancestral strain was observed during in vitro studies [15,16]. Following this, in mid-April 2021, we decided to track breakthrough COVID-19 infections across the country and gain cognizance of the different variants that were responsible for such infections.

In April 2021, Hacisuleyman et al. reported 417 cases of breakthrough infections in individuals vaccinated with Pfizer and Moderna messenger ribonucleic RNA (mRNA) vaccines [17]. Further, post-vaccination breakthrough COVID-19 infections are being reported from all over the globe. The Centers for Disease Control and Prevention (CDC) reported a total of 10,262 COVID-19 vaccine breakthrough infections till April 2021 [18]. Breakthrough infections in healthcare workers vaccinated with the BNT162b2 vaccine were reported from Italy in the May 2021 during an outbreak of SARS-CoV-2 lineage B.1.1.7 [19]. In India, a few studies reported breakthrough infections in small parts of the country, such as in Kerala [20], Chennai [21], and Delhi [22]. Taking cognizance of such reports, in April–May 2021, a nationwide study was undertaken to understand the clinico-demographic profile of patients and SARS-CoV-2 strains responsible for post-vaccination breakthrough COVID-19 infections across the country. To our understanding, this is the largest and first nationwide study of post-vaccination breakthrough infections in India.

2. Materials and Methods

2.1. *Definition of COVID-19 Breakthrough Infection*

A breakthrough COVID-19 infection was defined as an individual testing positive for SARS-CoV-2 by rRT-PCR or rapid antigen test (RAT) any time after 14 days of receiving one dose of any of the licensed COVID-19 vaccines.

2.2. *Study Catchment Area*

The Indian Council of Medical Research's Department of Health Research (ICMR-DHR) utilized a network of viral research and diagnostic laboratories (VRDLs) to track breakthrough infections. Clinical and demographic details as well as nasopharyngeal/oropharyngeal swabs (NPS/OPS) of COVID-19 patients satisfying the case definition were collected by the VRDLs in the north, south, west, east, northeast, and central parts of India from 17 states and Union Territories (UTs) (Maharashtra, Kerala, Gujarat, Uttarakhand, Karnataka, Manipur, Assam, Jammu and Kashmir, Chandigarh, Rajasthan, Madhya Pradesh, Punjab, Pondicherry, New Delhi, West Bengal, Tamil Nadu, and New Delhi) from 5 March 2021 till 3 June 2021. These clinical specimens were sequenced using next-generation sequencing (NGS) to determine nucleotide variations in the SARS-CoV-2 genome from the identified viral strains.

2.3. *Inclusion, Exclusion Criteria and Transport of Specimens*

Cases fulfilling the case definition and the following inclusion criteria were enrolled in the study: (i) cases with or without previous history of COVID-19; (ii) cases whose real-time RT-PCR threshold value was <30 and NPS/OPS were appropriately stored at −80 °C; and (iii) sample referral forms (SRF) capturing the demographic and clinical details of cases were available with the respective VRDLs. All the specimens of breakthrough cases fulfilling the above criteria were packed in triple-layer packaging with dry ice as per International Air Transport Association (IATA) guidelines and then transferred to the reference laboratory at the ICMR-National Institute of Virology (ICMR-NIV), Pune, for sequencing and variant analysis. As depicted in Figure 1, on 8 April 2021, a new specimen referral form (SRF) that included details of COVID-19 vaccination was launched by ICMR throughout the country to capture information related to vaccination status at the time of COVID-19 testing. However, quite a few states had not implemented this new SRF.

Therefore, it was not possible to track COVID-19 breakthrough infections in these states, and they were not included in this study.

2.4. Retrieval of Clinical and Demographic Data

Though completely filled SRFs were requested along with the specimens, most of the forms received from laboratories were incomplete due to the increased burden of testing during the second wave of COVID-19 in India. Therefore, telephonic interviews were conducted, wherein each reported breakthrough case was called and interviewed individually during the period of 25 May to 14 June 2021. The telephonic interviews also helped in validating the information available in the SRF and in filling in missing data. The patients were questioned on their demographic details, vaccination status, history of earlier COVID-19 infection, contact with a laboratory-confirmed case of COVID-19 prior to breakthrough infection, presence of comorbidities, symptoms developed, and course of infection, including the details of hospitalization. Each phone call typically lasted for 10–12 min, and only patients who provided a complete history were included in the study. A total of 814 clinical specimens were received from the different VRDLs all over the country at ICMR-NIV, Pune. The onset date/OPS and NPS collection dates ranged from 5 March to 3 June 2021, which coincided with the second wave of the COVID-19 pandemic in India. Out of these, 15 patients were not vaccinated for COVID-19, while 2 patients did not give any vaccination history, and 120 patients could not be traced. Thus, after excluding these 137 patients, a total of 677 cases were included in the study. A total of 10 of these 677 patients had documented COVID-19 infection between 8 and 14 days after receiving one dose of vaccine. Though these 10 cases did not satisfy the case definition, they were included in the study, as we did not want to lose any opportunity to detect the newly identified SARS-CoV-2 variants AY.1 and AY.2.

2.5. RNA Extraction and Next Generation Sequencing

Total RNA was extracted from 200 to 400 µL of NPS/OPS samples using an automated RNA extraction system (Thermo Fisher, Waltham, MA, USA) using Magmax RNA extraction kit (Applied Biosystems, Waltham, MA, USA). Real time RT-PCR (reverse transcriptase polymerase chain reaction) was set using SARS-CoV-2-specific primers for the detection of E gene and RdRP (RNA-dependent RNA polymerase) gene as described earlier [23]. The IlluminaCovidseq protocol (IlluminaInc, San Diego, CA, USA) was followed for preparation of RNA libraries. Extracted RNA was annealed using random hexamers to prepare for cDNA (complementary DNA) synthesis. The first strand of cDNA was synthesized using reverse transcription. The synthesized cDNA was amplified in two separate PCR plates using two pools of primers (pool 1 and pool 2) covering the entire genome of SARS-CoV-2. Amplified cDNA was then tagmented and bead-based post-tagmentation clean-up was performed. Tagmented amplicons were further amplified in this step using a PCR program as per manufacturer's instructions (Covidseq reference guide, Illumina). This PCR step added pre-paired 10 base pair indexes (Set 1, 2, 3, 4 adapters), required for sequencing cluster generation. One Covidseq positive control (CPC) and one negative template control (NTC) were used for each 96-well plate. Libraries generated in batches of 96 samples per plate were pooled into one 1.7 mL tube. Libraries of optimal size were purified by using a magnetic bead-based cleanup process method. Amplified and purified libraries were quantified using a KAPA Library Quantification Kit (KapaBiosystems, Roche Diagnostics Corporation, Indianapolis, IN, USA).

For a set of 384 samples, 25 µL of each normalized pool containing index adapter set 1, 2, 3, 4 was combined in a new micro-centrifuge tube. At this step, a final pool of 384 samples was diluted to a starting concentration of 4 nM. These libraries were then denatured, diluted, and then loaded at a final loading concentration of 1.4pM onto the NextSeq 500/550 system using NextSeq 500/550 High Output Kit v2.5 (75 Cycles) as per the manufacturer's instructions (Illumina Inc., San Diego, CA, USA). The files were analyzed using the reference-based mapping method, as implemented in CLC genomics workbench version 20.0 (CLC, QIAGEN, Aarhus, Denmark). The Wuhan Hu-1 isolate (Accession Number: NC_045512.2) was used as the reference sequence to retrieve the genomic sequence of the SARS-CoV-2. The retrieved sequences were aligned, along with few representative sequences from the GISAID database, in CLC Genomics Workbench v.20. A phylogenetic tree was generated using the MEGA software [24] for the aligned sequences. Gene-wise amino acid mutations were also observed.

3. Results

3.1. Clinical and Demographic Analysis of the Breakthrough Samples

Detailed distribution of breakthrough cases (n = 677) collected from 17 states/UT of the country used for NGS is provided in Table 1. The clinical samples for analysis were collected between March and June 2021. Out of these 677 patients, 85 acquired COVID-19 after taking the first dose of the vaccine, while 592 were infected after receiving both doses of the vaccine. A total of 517 of these 592 individuals contracted COVID-19 after 2 weeks of receiving the second dose of vaccine.

Table 1. Region-wise and state-wise distribution of SARS-CoV-2 clinical samples used for next-generation sequencing (n = 677).

Region	State/UTs	Clinical Samples Received from Each Site
North India	New Delhi	20
	Uttarakhand	50
	Jammu and Kashmir	25
	Punjab	12
	Chandigarh	19
Northeastern India	Manipur	15
	Assam	40
Eastern India	West Bengal	10
	Jharkhand	12
Central India	Madhya Pradesh	68
Western India	Maharashtra	53
	Gujarat	47
	Rajasthan	58
South India	Karnataka	181
	Kerala	16
	Tamil Nadu	25
	Puducherry	26

Clinical samples from the COVID-19 cases post second dose of vaccination were collected with a median of 38 days and had an inter-quartile range (IQR) of 20 (19–58) days. A total of 604 patients had received Covishield/AstraZeneca vaccine, 71 had received Covaxin, and 2 had received Sinopharm vaccine (BBIBP-CorV).

Clinical data were analyzed for 677 breakthrough cases. The median age (and the IQR) of patients in the study was 44 (31–56); of the breakthrough cases after one dose, the median age was 53 (45–61), and after two doses it was 41 (30–55). A total of 441 (65.1%) of the breakthrough cases were males. A total of 482 cases (71%) were symptomatic with one or more symptoms, while 29% had asymptomatic SARS-CoV-2 infection. Fever (69%) was the most consistent presentation, followed by body ache, including headache and nausea (56%), cough (45%), sore throat (37%), loss of smell and taste (22%), diarrhea

(6%), and breathlessness (6%), and 1% had ocular irritation and redness. The clinical and demographic analysis of the 677 cases of breakthrough infections is enumerated in Table 2.

Table 2. Demographic analysis of breakthrough COVID-19 infections.

Characteristics	Vaccinated with Both Doses *	Vaccinated with One Dose *	Total Cases
	(*N* = 592)	(*N* = 85)	(*N* = 677)
	n (% of Total)	*n* (% of Total)	*n* (% of Total)
Age (Years)			
Median (Interquartile range)	41(30–55)	53 (45–61)	44 (31–56)
Gender			
Male	383 (64.7)	58 (68.2)	441 (65.1)
Female	209 (35.3)	27 (31.8)	236 (34.9)
Other	NIL	NIL	NIL
Comorbidities			
Yes	134 (22.6)	20 (23.5)	154 (22.7)
No	458 (77.4)	65 (76.5)	523 (77.3)
Missing	NIL	NIL	NIL
Type of Vaccine			
Covaxin	63 (10.64)	8 (9.4)	71 (10.5)
Covishield/AstraZeneca	527 (89.02)	77 (90.6)	604 (89.2)
Sinopharm	2 (0.33)	0	2 (0.3)
Contact with lab-confirmed SARS-CoV-2 case	282 (47.6)	31 (36.5)	313 (46.2)
Median gap between 2 doses of the vaccine(days)	33(29–41)	NA	NA
Median (IQR) interval in days between vaccination and SARS-CoV-2 test	39(19–58)	26(18–38)	NA
Symptoms during the course of illness			
Yes	411 (69.4)	71 (83.5)	482 (71.2)
No	181 (30.6)	14 (16.5)	195 (28.8)
Hospitalized	53 (8.9)	14 (16.5)	67 (9.9)
Individuals with comorbidities #	22 (3.7)	8 (9.4)	30 (4.4)
Individuals without comorbidities #	31 (5.2)	6 (7.1)	37 (5.5)
Clinical outcome			
Alive	589(99.5)	85 (100)	674 (99.6)
Dead	3 (0.5)	0	3 (0.4)

* $p = 0.0086$ and odds ratio = 0.44 for the proportions with symptomatic among vaccinated with two doses and vaccinated with one dose. # $p = 0.0328$ and Odds ratio = 0.49 for the proportions with individuals hospitalized with comorbidities and individuals hospitalized with non-comorbidities.

Comorbidities were observed in the 154 out of 677 cases, which included diabetes mellitus type 2 and hypertension as well as chronic cardiac, renal, and pulmonary diseases and obesity. The symptoms reported in patients with breakthrough infections are enumerated in Figure 2. The cases with comorbidities were significantly predisposed to develop symptoms (cough, sore throat, fever, loss of smell and taste, diarrhea, breathlessness, ocular symptoms, and constitutional symptoms (body ache, headache, nausea)); (OR = 2.0042, 95% C.I. = 1.2857 to 3.1244, *z*-statistic = 3.069, $p = 0.0021$). Additionally, the cases with medical comorbidities were significantly more predisposed to hospitalization (OR = 3.1779, 95% C.I. = 1.8886 to 5.3471, z-statistic = 4.355, $p < 0.0001$).

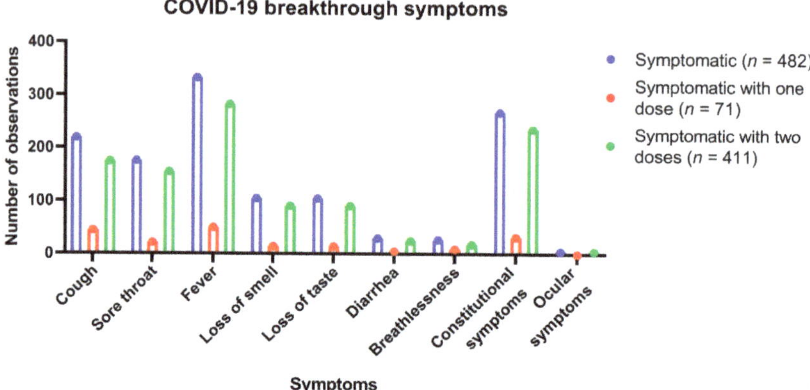

Figure 2. Symptoms reported in COVID-19 breakthrough infections.

3.2. Vaccine Breakthrough Infections in Individuals with Previous History of COVID-19

Twelve vaccinated individuals gave definitive history of previous laboratory confirmed COVID-19 infection. All of them subsequently received two doses of Covishield/AstraZeneca. The Indian Council of Medical Research had earlier conducted a study that defined re-infection as positive test for SARS-CoV-2 on two separate occasions by either molecular or rapid antigen test at an interval of at least 102 days, with one negative molecular test in between [25]. While we could not elicit history of a negative test result following the first episode of COVID-19 infection, the gap between two positive tests was well above 102 days in 11 cases. One individual received his first dose of COVID-19 vaccine 40 days after testing positive for COVID-19. He received dose two of Covishield/AstraZeneca after 5 weeks, and tested positive 15 days after receiving the second dose. Though the time period between him testing positive twice for SARS-CoV-2 was less than 102 days (89 days), this case was included in the analysis because, to the best of our knowledge, the literature shows that the maximal duration of SARS-CoV-2 shedding that is detectable by PCR is 63 days after onset of symptoms [26]. Hence, we considered this case as a true re-infection and not mere shedding of genomic RNA. This individual was asymptomatic. Median duration from first bout of infection to first dose of vaccination in these 12 cases was 135 days (IQR = 85–166.75 days). Median gap between two doses of the vaccine was 32 days (IQR = 29.75–38 days). Median duration of breakthrough infection from the second dose of the vaccine was 45 days (IQR = 17–55.5 days) and between earlier COVID-19 infection (day of testing) and breakthrough COVID-19 infection (day of testing) was 196 days (IQR = 177.5–249.25 days). Six of these individuals were symptomatic. Most commonly reported symptoms were body ache (4/6), fever (3/6) cough (2/6), sore throat (2/6), headache (2/6), chest pain (1/6). A total of 4 patients had comorbidities, and 1 person out of 12 required hospital admission. He was symptomatic (cough, cold, fever) but had no associated comorbidities.

3.3. Next-Generation Sequencing Analysis of the Breakthrough Specimens

Out of 677 cases included in this study, sequencing was not performed for 112 clinical samples (two doses: $n = 95$; single dose: $n = 17$) based on the higher Ct and low Kappa value.

The complete genome of 511 SARS-CoV-2 were recovered with genome coverage of more than 98% (two doses: $n = 446$; single dose: $n = 65$). SARS-CoV-2 sequences with more than 99% and 84% of the genome coverage were recovered from 446 (two doses: $n = 387$;

single dose: n = 59) and 546 (two doses: n = 480; single dose: n = 66) clinical specimens, respectively. Less than 98% of genomes were retrieved from 54 samples and were not used further in analysis. The details of the percentage genome retrieved, total reads mapped, and percentage relevant reads are given in Supplementary Table S1. The lineages were retrieved using the Pangolin online software (https://cov-lineages.org/pangolin.html; accessed on 8 August 2021) from the specimens with more than 84% genome coverage and mentioned in Supplementary Table S1.

The geographic distribution of the different SARS-CoV-2 variants with 98% genome coverage were characterized using Pangolin software and are presented in Figure 3. Delta (B.1.617.2) (n = 384) was the major SARS-CoV-2 lineage observed in the breakthrough samples after two doses of vaccine, followed by alpha (B.1.1.7) (n = 28). Kappa (B.1.617.1) (n = 22), B.1.617.3 (n = 2), B (n = 1), B.1.36 (n = 5), B.1.1.294 (n = 1), B.1.36.16 (n = 1), B.1.1.306 (n = 1), and Delta AY.2 (n = 1) pangolin lineage variants were also observed along with others; details are given in Supplementary Table S1. A total of 65 out of 85 samples from individuals infected with SARS-CoV-2 after one dose of vaccination had 99.5% genome retrieval. These sequences had Delta (B.1.617.2) (n = 59), Alpha (B.1.1.7) (n = 4) Kappa (B.1.617.1) (n = 1), and Delta AY.1 (n = 1). The Delta AY.1 variant was observed in Madhya Pradesh (MP), while Delta AY.2 was observed in the Rajasthan state of India. The percentage nucleotide divergence of the different SARS-CoV-2 strains relative to reference was 99.81–100%; details for each strain are given in Supplementary Table S1.

It was observed that southern, western, eastern and northwestern regions of India predominantly reported breakthrough infections from mainly Delta and then Kappa variant of SARS-CoV-2. The northern and central regions reported such infections due to Alpha, Delta, and Kappa variants; however, cases due to Alpha variant predominated in the northern region (Figure 3). The overall majority (86.09%) of the breakthrough infections were caused by the Delta variant (B.1.617.2) of SARS-CoV-2 in different regions of India, except for the northern region where the Alpha variant predominated.

Of the 12 cases of breakthrough infection with previous history of COVID-19, 6 samples could be sequenced. These sequences included Delta (B.1.617.2) (n = 1), B.1.1.7 (n = 2), Kappa (B.1.617.1) (n = 1), and B.1.36 (n = 2). B.1.1.7 was sequenced from the individual who tested positive for SARS-CoV-2 twice at an interval of 89 days.

Figure 4 depicts the neighbor-joining tree generated using the Tamura-3-parameter model with a bootstrap replication of 1000 cycles. SARS-CoV-2 sequences (n = 421) with genome coverage of 99% and fewer gaps in coding regions were taken for the generation of a phylogenetic tree. A total of 32 representative and 421 SARS-CoV-2 sequences retrieved in this study were used to generate the phylogenetic tree. The Delta sequences (n = 358) represented the highest proportion of breakthrough cases from different parts of the country and clustered into four distinct sub-lineages. Sub-lineage I had 166 SARS-CoV-2 sequences, while sub-lineages II, III, and IV had 100, 68, and 24 sequences, respectively, which are marked on the phylogenetic tree. The gene-wise amino acid mutations were further looked upon for the retrieved sequences and the representative sequences relative to the reference sequence. It was observed that the Delta SARS-CoV-2 variant sequences had conservation in different gene positions, leading to differential clustering. These conserved mutations of different sub-lineages are depicted in Figure 5. Sub-lineage I (red color): mutations in ORF1ab A1306S, P2046L, P2287S, V2930L, T3255I, T3446A, G5063S, P5401L, and A6319V and in N G215C; **Sub-lineage II (green color):** ORF1ab P309L, A3209V, V3718A, G5063S, and P5401L and ORF7a L116F; **Sub-lineage III (pink color):** ORF1ab A3209V, V3718A, T3750I, G5063S, and P5401L and spike A222V; **Sub-lineage IV (Orange color):** ORF1ab P309L, D2980N, and F3138S and spike K77T. **Common in B.1.617.2 lineage:** ORF1ab P4715L; spikeT19R, L452R, T478K, D614G, and P681R; ORF3a S26L; M I82T; ORF7a V82A and T120I; and N D63G, R203M, and D377Y.

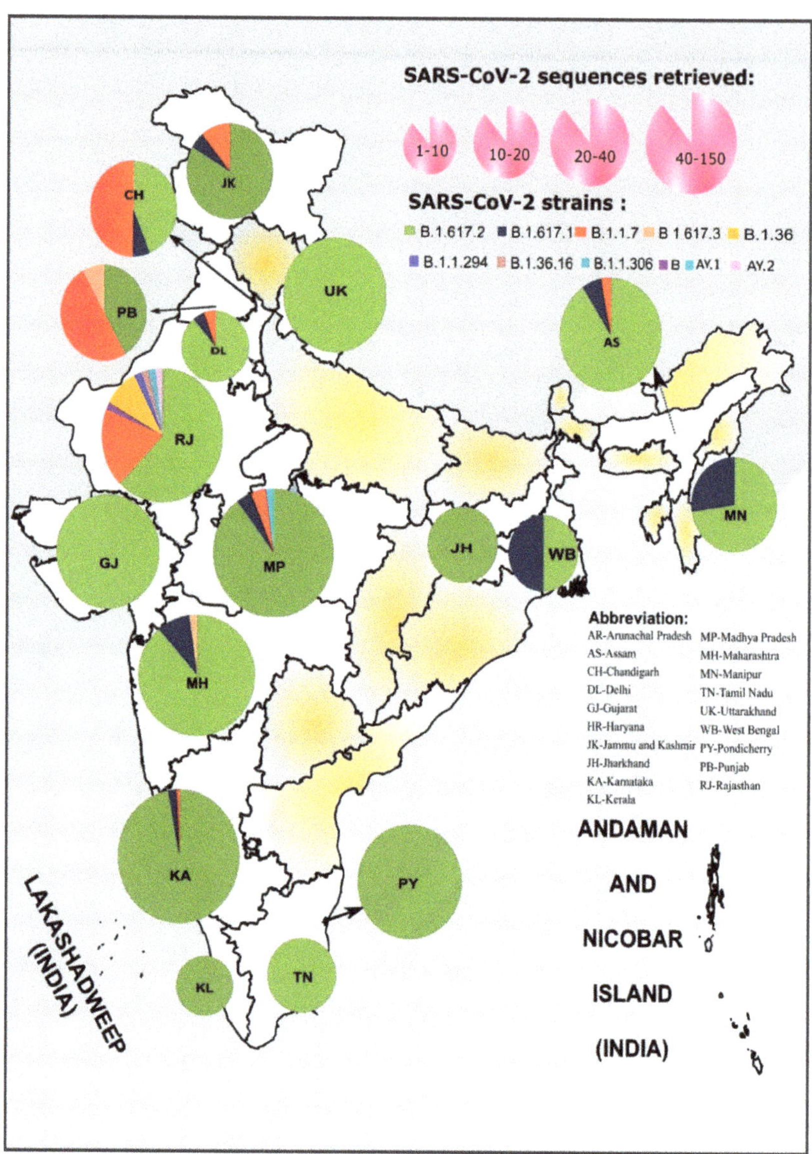

Figure 3. Distribution of the SARS-CoV-2 genome prevalence among cases of breakthrough infection. The size of each pie chart within the states of the India map is ranged based on the number of sequences retrieved in the study. The distribution in the pie chart is proportional to the numbers in each respective clade in each state. The outline of India's map was downloaded from http://www.surveyofindia.gov.in/file/Map%20f%20India_1.jpg (accessed on 20 March 2020) and further modified to include relevant data in the SVG editor.

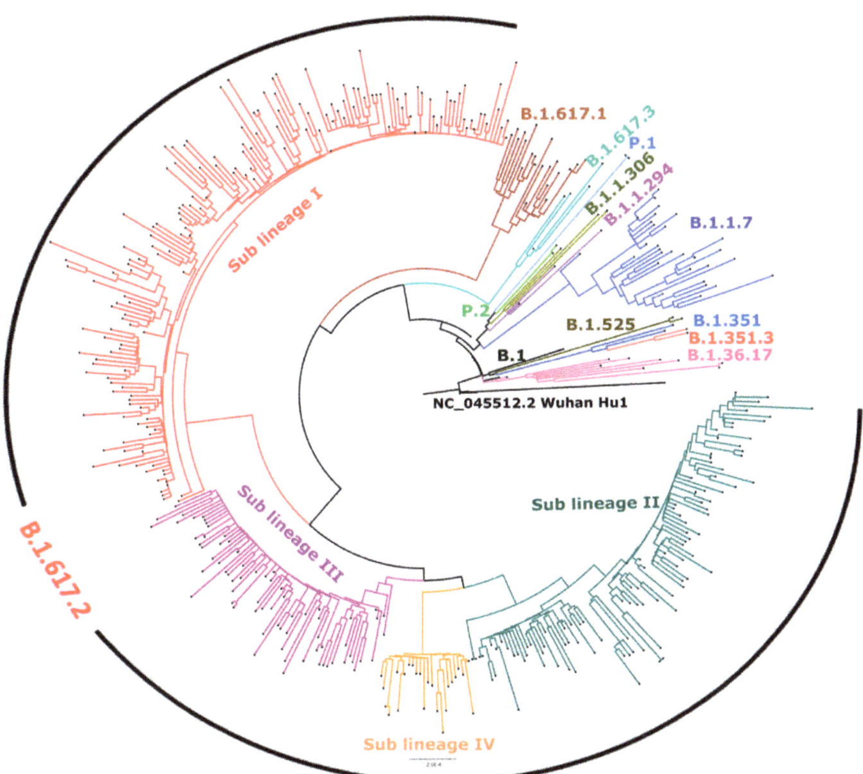

Figure 4. Phylogenetic tree of the 402 SARS-CoV-2 genomes from breakthrough cases with one and two doses vaccine recipients. A Neighbor-joining tree of the 402 SARS-CoV-2 sequences retrieved in this study, along with the representative SARS-Cov-2 sequences from different clades with a bootstrap replication of 1000 cycles. Four major sub-lineages of Delta variant were observed, which are marked on branched in different colors. Sub-lineages I–IV are marked in red, green, pink, and orange color on the nodes, respectively. B.1.617.1 sequence is marked in brown and B.1.617.3 in blue color. The representative pangolin lineages are also marked on branches in different colors. FigTree v1.4.4 and Inkscape were used to visualize and edit the generated tree.

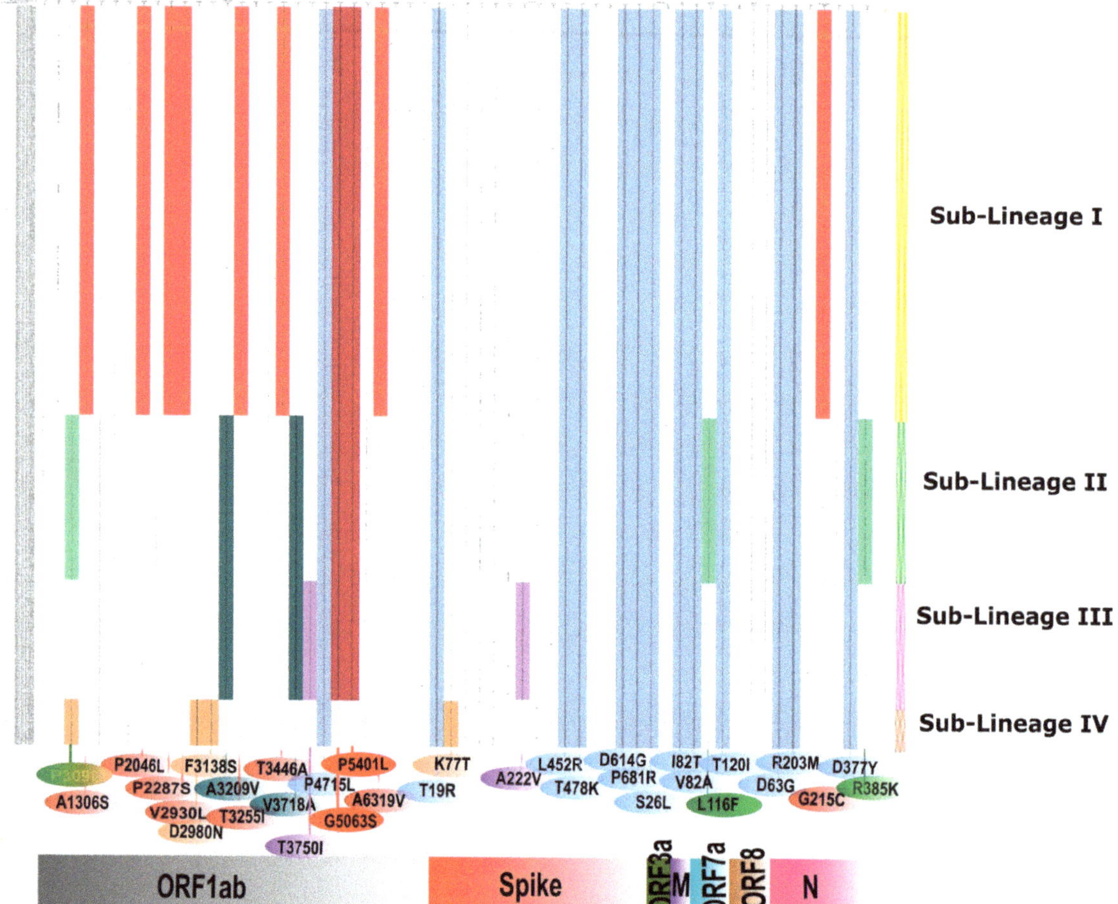

Figure 5. Characterization of sub-lineages observed in the Delta SARS-CoV-2 variant from breakthrough sequences: Sub-lineages I–IV are marked in red, green, pink, and orange color along *y* axis. The amino acid mutations observed in different genes are marked on the *x*-axis. The amino acid change is marked concerning Wuhan-HU-1 (Accession No: NC_045512.2). It is observed that a couple of mutations are conserved in sub-lineages I-III and marked as the gradient of red-violet color. Amino acid mutations conserved for sub-lineage II and III are marked as a violet-green gradient. The amino acid changes common to Delta variant is marked in blue color.

4. Discussion

Globally, COVID-19 vaccines were accorded Emergency Use Authorization (EUA) and introduced quickly into public health programs to prevent SARS-CoV-2 infections and curtail disease transmission, thus saving lives and livelihood. However, the emergence of SARS-CoV-2 VOCs has raised a public health concern due to increased transmissibility and potential to evade humoral immune response. The fact that this has happened amid vaccination uptake has created a dilemma about the efficacy of vaccines under EUA. Data show that there is a 3-fold and 16-fold reduction in neutralization against the Delta and Beta variants as compared with the Alpha variant with BNT162b2 vaccinated sera, and a 5-fold and 9-fold reduction against the same with ChAdOx1 nCoV-19 [27].

As per the WHO classification, the Delta variant has been designated as a variant of concern due to increased transmission and higher immune evasion, whereas the other two sub-lineages of B.1.617—namely, B.1.617.1 and B.1.617.3—with E484Q are grouped in

VUI [28]. The B.1.617 variant and its lineage B.1.617.2 were primarily responsible for the surge in COVID-19 cases in Maharashtra state [29]. Delta (B.1.617.2) and Kappa (B.1.617.1) were detected among 60% of the clinical specimens of the COVID-19 cases collected from Maharashtra during January and February 2021 [30]. The rapid rise in daily infections was observed in India, with dominance of the Delta variant, which accounted for >99% of all sequenced genomes in April 2021 [31].A recent study on the secondary attack rates in UK households demonstrated a higher transmission of Delta compared with the Alpha variant [8]. The reduced neutralizing capability of currently used SARS-CoV-2 vaccines against Delta variants is one of the causes for recent increases in breakthrough cases with this strain.

Emergence of VOCs has led to an upsurge in COVID-19 cases and a subsequent wave of pandemic in various countries including India. Incidentally, several countries have reported COVID-19 breakthrough infections even after completion of full vaccination schedules [17,19,30]. More than 10,000 breakthrough infections after completion of a full course of vaccination have been reported in the USA. Overall breakthrough infections were seen in a smaller percentage of the total vaccinated population [18]. A recent study has also reported mild symptomatic breakthrough infections from Kerala and Delhi, India [20,22].

The present study revealed that the infection among breakthrough cases predominantly occurred through the Delta variant, indicating its high community transmission during this period, followed by Alpha and Kappa variants. In our study, 67 cases (9.8%) required hospitalization, and fatality was observed in only 3 cases (0.4%). This clearly suggests that vaccination reduces the severity of disease, hospitalization, and mortality. Therefore, enhancing the vaccination drive and immunizing populations quickly would be the most important strategy for preventing further deadly waves of COVID-19 and would reduce the burden on the health care system.

When COVID-19 vaccination was launched in India on 16 January 2021, the recommended gap between two doses of the Covishield/AstraZeneca vaccine was 4 weeks [32]. Later on, based on studies from the Oxford Vaccine Group and the WHO interim recommendations, the dose interval was increased to 6–8 weeks [33,34] and then to 12–16 weeks within a small time frame. This was based on effectiveness data from the UK [35] and recommendations of Canada [36]. Since Covishield/AstraZeneca contributes to almost 90% of the vaccination in India, increased dose spacing has led to vaccination of greater numbers of eligible people with at least one vaccine dose. However, to tackle the Delta variant of SARS-CoV-2, the UK's Joint Committee on Vaccination and Immunisation (JCVI) recommended a shortening of the dosing interval to 8 weeks for priority cohorts who are at risk of COVID-19 [37]. Since the Delta variant was predominantly sequenced in our breakthrough infection cases during the second wave of COVID-19 in India, focused studies are now being commissioned in India to look at the need for reducing the gap between two doses of Covishield/AstraZeneca for specific population groups such as immunocompromised individuals, transplant recipients, cancer patients, and people living with HIV.

Two new SARS-CoV-2 variants, Delta AY.1 and AY.2, were also identified in these study samples. The AY.1 and AY.2 variants have been aggregated with Delta variant B.1.617.2 [12]. Delta AY.1 and AY.2 are characterized by the presence of the K417N mutation in the spike protein region. K417N, E484K, L452R, and E484Q are the mutations known to disrupt receptor-binding domain (RBD) binding capacity, making them more infectious by immune escape against the current vaccines [38]. This indicates improved viral fitness to evade immune responses and survive against the vaccines.

Post-vaccination breakthrough COVID-19 cases have been reported from various countries with the use of different licensed vaccines. It appears that the current COVID-19 vaccines are disease-modifying in nature, wherein mild or less severe infections are expected to occur in vaccinated individuals. However, vaccination seems to have an obvious advantage in averting severe disease, hospitalizations, and deaths. Therefore, continuous monitoring of post-vaccination breakthrough infections, along with monitoring of clinical severity of disease, must be adopted as an essential component of vaccine roll-out programs in all countries. Such monitoring will help us to understand the need for adequately tweaking the available vaccines and also for developing new vaccines with enhanced potential to protect against variant strains of SARS-CoV-2.

Identification of the new variants that is responsible for the breakthrough infections underline the importance of this study. It also highlights the need for active genomic surveillance of the new SARS-CoV-2 variants and for assessing their potential to evade immune responses.

Supplementary Materials: The following are available online at https://www.mdpi.com/article/10.3390/v13091782/s1, Table S1: Percentage genome retrieved, total reads mapped, and percentage relevant reads of the SARS-CoV-2 genomes retrieved in this study.

Author Contributions: Conceptualization, P.D.Y. and N.G.; methodology, P.D.Y., N.G., H.K., L.M., R.R.S., A.K., D.A.N., A.M.S., S.P., T.M., M.D., P.P. and Y.J.; software, D.A.N., A.K., M.D., P.P. and Y.J.; resources, S.R. (Salaj Ranaand), S.D. (Shanta Dutta), S.G., J.N., N.V., N.A., G.N., A.K.B., M.P.K., D.B., P.B. (Pradip Barde), J.I., S.R. (Sharmila Raut), S.D. (Sulochana Devi), P.B. (Purnima Barua), P.G., B.B., D.K. (Deepjyoti Kalita), K.D., B.F., K.G., R.J., A.M., R.D., S.D. (Sarada Devi), D.K. (Deepa Kinariwala), N.K., Y.K.T., P.K.K., A.G., H.K. (Himanshu Khatri), B.M., M.N., L.D., N.S. and J.S.; writing original draft preparation, P.D.Y., N.G. and D.A.N.; writing, P.D.Y., N.G., H.K. (Harmanmeet Kaur) and D.A.N. supervision, P.D.Y., N.G., R.R.S. and A.M.S.; project administration, P.D.Y., N.G. and P.A.; funding acquisition, P.D.Y. and P.A. All authors have read and agreed to the published version of the manuscript.

Funding: The study was conducted with intramural funding for 'Molecular epidemiological analysis of SARS-CoV-2 is circulating in different regions of India' of Indian Council of Medical Research (ICMR), New Delhi, provided to ICMR-National Institute of Virology, Pune.

Institutional Review Board Statement: The study was approved by the Institutional Human Ethics Committee of ICMR-NIV, Pune, India under project 'Molecular epidemiological analysis of SARS-CoV-2 circulating in different regions of India'.

Informed Consent Statement: Informed consent was obtained from all subjects involved in the study.

Data Availability Statement: All the sequencing data and information of this study is available in GISAID. Accession no is provided in supplementary table.

Acknowledgments: Authors gratefully acknowledge the encouragement and support extended by Balram Bhargava, Secretary to the Government of India, Department of Health Research, Ministry of Health and Family Welfare and Director-General, ICMR and Samiran Panda, ECD Chief, ICMR, Delhi. We thank the team members of the Maximum Containment Facility, ICMR-NIV, Pune, including Deepak Patil, Pranita Gawande, Kaumudi Kalele, Ashwini Waghmare, Tejashri Kore, Shilpa Ray, Priyanka Waghmare, and Poonam Bodke for excellent support. The authors would like to acknowledge Krishnapal Karmodia and Sanjeev Galande from IISER, Pune, for helping us utilize their NGS facility. Further, we also acknowledge Anjani Gopal from IISER, Pune, for assisting us in the NGS facility. We would like to acknowledge all the authors that have submitted the SARS-CoV-2 sequences to the GISAID database.

Conflicts of Interest: Authors do not have conflicts of interest.

References

1. Zhang, L.; Jackson, C.B.; Mou, H.; Ojha, A.; Rangarajan, E.S.; Izard, T.; Farzan, M.; Choe, H. SARS-CoV-2 spike-protein D614G mutation increases virion spike density and infectivity. *Nat. Commun.* **2020**, *11*, 6013. [CrossRef] [PubMed]
2. Yadav, P.D.; Nyayanit, D.A.; Sahay, R.R.; Sarkale, P.; Pethani, J.; Patil, S.; Baradkar, S.; Potdar, V.; Patil, D.Y. Isolation and Characterization of the New SARS-CoV-2 Variant in Travellers from the United Kingdom to India: VUI-202012/01 of the B.1.1.7 Lineage. *J. Travel Med.* **2021**, *28*, taab009. [CrossRef] [PubMed]
3. Sapkal, G.; Yadav, P.D.; Ella, R.; Abraham, P.; Patil, D.Y.; Gupta, N.; Panda, S.; Mohan, V.K.; Bhargava, B. Neutralization of B.1.1.28 P2 Variant with Sera of Natural SARS-CoV-2 Infection and Recipients of Inactivated COVID-19 Vaccine Covaxin. *J. Travel Med.* **2021**. Epub ahead of print.

4. Yadav, P.D.; Gupta, N.; Nyayanit, D.A.; Sahay, R.R.; Shete, A.M.; Majumdar, T.; Patil, S.; Kaur, H.; Nikam, C.; Pethani, J.; et al. Imported SARS-CoV-2 V501Y.V2 Variant (B.1.351) Detected in Travelers from South Africa and Tanzania to India. *Travel Med. Infect. Dis.* **2021**, *41*, 102023. [CrossRef] [PubMed]
5. Yadav, P.D.; Nyayanit, D.A.; Majumdar, T.; Patil, S.; Kaur, H.; Gupta, N.; Shete, A.M.; Pandit, P.; Kumar, A.; Aggarwal, N.; et al. An Epidemiological Analysis of SARS-CoV-2 Genomic Sequences from Different Regions of India. *Viruses* **2021**, *13*, 925. [CrossRef] [PubMed]
6. Rambaut, A.; Holmes, E.C.; O'Toole, Á.; Hill, V.; McCrone, J.T.; Ruis, C.; du Plessis, L.; Pybus, O.G. A Dynamic Nomenclature Proposal for SARS-CoV-2 Lineages to Assist Genomic Epidemiology. *Nat. Microbiol.* **2020**, *5*, 1403–1407. [CrossRef]
7. Tegally, H.; Wilkinson, E.; Lessells, R.J.; Giandhari, J.; Pillay, S.; Msomi, N.; Mlisana, K.; Bhiman, J.N.; von Gottberg, A.; Walaza, S.; et al. Sixteen Novel Lineages of SARS-CoV-2 in South Africa. *Nat. Med.* **2021**, *27*, 440–446. [CrossRef]
8. Faria, N.R.; Mellan, T.A.; Whittaker, C.; Claro, I.M.; Candido, D.D.S.; Mishra, S.; Crispim, M.A.E.; Sales, F.C.; Hawryluk, I.; McCrone, J.T.; et al. A variant lineage of SARS-CoV-2 associated with rapid transmission in Manaus, Brazil, evolved in November 2020 with immune escape characteristics. *Science* **2021**, *372*, 815–821. [CrossRef]
9. Yadav, P.D.; Sapkal, G.N.; Abraham, P.; Deshpande, G.; Nyayanit, D.A.; Patil, D.Y.; Gupta, N.; Sahay, R.R.; Shete, A.M.; Kumar, S.; et al. Neutralization Potential of Covishield Vaccinated Individuals Sera against B.1.617.1. *Clin. Infect. Dis.* **2021**, ciab483. [CrossRef]
10. Yadav, P.D.; Sapkal, G.N.; Ella, R.; Sahay, R.R.; Nyayanit, D.A.; Patil, D.Y.; Deshpande, G.; Shete, A.M.; Gupta, N.; Mohan, V.K.; et al. Neutralization of Beta and Delta variant with sera of COVID-19 recovered cases and vaccinees of inactivated COVID-19 vaccine BBV152/Covaxin. *J. Travel Med.* **2021**, taab104. [CrossRef]
11. Cherian, S.; Potdar, V.; Jadhav, S.; Yadav, P.; Gupta, N.; Das, M.; Rakshit, P.; Singh, S.; Abraham, P.; Panda, S.; et al. SARS-CoV-2 Spike Mutations, L452R, T478K, E484Q and P681R, in the Second Wave of COVID-19 in Maharashtra, India. *Microorganisms* **2021**, *9*, 1542. [CrossRef] [PubMed]
12. Available online: https://www.cdc.gov/csels/dls/locs/2021/07-06-2021-lab-advisory-SARS-CoV2_Variants_AY_1_and_AY_2_Now_Aggregated_with_Delta_Variant_B_1_617_2.html (accessed on 31 August 2021).
13. CoWIN Dashboard. Available online: https://dashboard.cowin.gov.in/ (accessed on 6 July 2021).
14. Kustin, T.; Harel, N.; Finkel, U.; Perchik, S.; Harari, S.; Tahor, M.; Caspi, I.; Levy, R.; Leshchinsky, M.; Ken Dror, S.; et al. Evidence for Increased Breakthrough Rates of SARS-CoV-2 Variants of Concern in BNT162b2-MRNA-Vaccinated Individuals. *Nat. Med.* **2021**, *27*, 1379–1384. [CrossRef] [PubMed]
15. Emary, K.R.W.; Golubchik, T.; Aley, P.K.; Ariani, C.V.; Angus, B.; Bibi, S.; Blane, B.; Bonsall, D.; Cicconi, P.; Charlton, S.; et al. Efficacy of ChAdOx1 NCoV-19 (AZD1222) Vaccine against SARS-CoV-2 Variant of Concern 202012/01 (B.1.1.7): An Exploratory Analysis of a Randomised Controlled Trial. *Lancet Lond. Engl.* **2021**, *397*, 1351–1362. [CrossRef]
16. Supasa, P.; Zhou, D.; Dejnirattisai, W.; Liu, C.; Mentzer, A.J.; Ginn, H.M.; Zhao, Y.; Duyvesteyn, H.M.E.; Nutalai, R.; Tuekprakhon, A.; et al. Reduced Neutralization of SARS-CoV-2 B.1.1.7 Variant by Convalescent and Vaccine Sera. *Cell* **2021**, *184*, 2201–2211.e7. [CrossRef]
17. Hacisuleyman, E.; Hale, C.; Saito, Y.; Blachere, N.E.; Bergh, M.; Conlon, E.G.; Schaefer-Babajew, D.J.; DaSilva, J.; Muecksch, F.; Gaebler, C.; et al. Vaccine Breakthrough Infections with SARS-CoV-2 Variants. *N. Engl. J. Med.* **2021**, *384*, 2212–2218. [CrossRef]
18. COVID-19 Vaccine Breakthrough Case Investigations Team; Birhane, M.; Bressler, S.; Chang, G.; Clark, T.; Dorough, L.; Fischer, M.; Watkins, L.F.; Goldstein, J.M.; Kugeler, K.; et al. COVID-19 Vaccine Breakthrough Infections Reported to CDC—United States, January 1–April 30, 2021. *MMWR Morb. Mortal. Wkly. Rep.* **2021**, *70*, 792–793. [CrossRef]
19. Loconsole, D.; Sallustio, A.; Accogli, M.; Leaci, A.; Sanguedolce, A.; Parisi, A.; Chironna, M. Investigation of an Outbreak of Symptomatic SARS-CoV-2 VOC 202012/01-Lineage B.1.1.7 Infection in Healthcare Workers, Italy. *Clin. Microbiol. Infect. Off. Publ. Eur. Soc. Clin. Microbiol. Infect. Dis.* **2021**, *27*, 1174.e1–1174.e4. [CrossRef]
20. Philomina, J.B.; Jolly, B.; John, N.; Bhoyar, R.C.; Majeed, N.; Senthivel, V.; Cp, F.; Rophina, M.; Vasudevan, B.; Imran, M.; et al. Genomic Survey of SARS-CoV-2 Vaccine Breakthrough Infections in Healthcare Workers from Kerala, India. *J. Infect.* **2021**, *83*, 237–279. [CrossRef] [PubMed]
21. Thangaraj, J.; Yadav, P.; Kumar, C.G.; Shete, A.; Nyayanit, D.A.; Rani, D.S.; Kumar, A.; Kumar, M.S.; Sabarinathan, R.; Kumar, V.S.; et al. Predominance of delta variant among the COVID-19 vaccinated and unvaccinated individuals, India, May 2021. *J. Infect.* **2021**, 2, 23. [CrossRef]
22. Tyagi, K.; Ghosh, A.; Nair, D.; Dutta, K.; Singh Bhandari, P.; Ahmed Ansari, I.; Misra, A. Breakthrough COVID19 Infections after Vaccinations in Healthcare and Other Workers in a Chronic Care Medical Facility in New Delhi, India. *Diabetes Metab. Syndr. Clin. Res. Rev.* **2021**, *15*, 1007–1008. [CrossRef] [PubMed]
23. Choudhary, M.L.; Vipat, V.; Jadhav, S.; Basu, A.; Cherian, S.; Abraham, P.; Potdar, V.A. Development of in Vitro Transcribed RNA as Positive Control for Laboratory Diagnosis of SARS-CoV-2 in India. *Indian J. Med. Res.* **2020**, *151*, 251–254. [CrossRef]
24. Kumar, S.; Stecher, G.; Li, M.; Knyaz, C.; Tamura, K. MEGA X: Molecular Evolutionary Genetics Analysis across Computing Platforms. *Mol. Biol. Evol.* **2018**, *35*, 1547–1549. [CrossRef] [PubMed]
25. Mukherjee, A.; Anand, T.; Agarwal, A.; Singh, H.; Chatterjee, P.; Narayan, J.; Rana, S.; Gupta, N.; Bhargava, B.; Panda, S. SARS-CoV-2 re-infection: Development of an epidemiological definition from India. *Epidemiol. Infect.* **2021**, *149*, e82. [CrossRef]
26. Cantini, F.; Niccoli, L.; Matarrese, D.; Nicastri, E.; Stobbione, P.; Goletti, D. Baricitinib therapy in COVID-19: A pilot study on safety and clinical impact. *J. Infect.* **2020**, *81*, 318–356. [CrossRef]

27. Abdool Karim, S.S.; de Oliveira, T. New SARS-CoV-2 Variants—Clinical, Public Health, and Vaccine Implications. *N. Engl. J. Med.* **2021**, *384*, 1866–1868. [CrossRef] [PubMed]
28. Salvatore, M.; Bhattacharyya, R.; Purkayastha, S.; Zimmermann, L.; Ray, D.; Hazra, A.; Kleinsasser, M.; Mellan, T.; Whittaker, C.; Flaxman, S.; et al. Resurgence of SARS-CoV-2 in India: Potential Role of the B.1.617.2 (Delta) Variant and Delayed Interventions. *medRxiv* **2021**. [CrossRef]
29. *Maharashtra: Double Mutant Found in 61% Samples Tested*; Indian Express: Mumbai, India, 2021.
30. Mullen, J.L.; Tsueng, G.; Latif, A.A.; Alkuzweny, M.; Cano, M.; Haag, E.; Zhou, J.; Zeller, M.; Hufbauer, E.; Matteson, N.; et al. Center for Viral Systems Biology. 2020. Available online: https://outbreak.info (accessed on 30 August 2021).
31. Keehner, J.; Horton, L.E.; Pfeffer, M.A.; Longhurst, C.A.; Schooley, R.T.; Currier, J.S.; Abeles, S.R.; Torriani, F.J. SARS-CoV-2 Infection after Vaccination in Health Care Workers in California. *N. Engl. J. Med.* **2021**, *384*, 1774–1775. [CrossRef]
32. Precautions and Contraindications for COVID-19 Vaccination. Available online: https://www.mohfw.gov.in/pdf/LetterfromAddlSecyMoHFWregContraindicationsandFactsheetforCOVID19vaccines.PDF (accessed on 23 August 2021).
33. Protection Enhanced if the Second Dose of COVISHIELD Is Administered between 6–8 Weeks. Available online: https://pib.gov.in/PressReleasePage.aspx?PRID=1706597 (accessed on 23 August 2021).
34. World Health Organization. Interim Recommendations for Use of the AZD1222 (ChAdOx1-S [Recombinant]) Vaccine against COVID19 Developed by Oxford University and AstraZeneca, 10 February 2021. World Health Organization. Available online: https://apps.who.int/iris/bitstream/handle/10665/339477/WHO-2019-nCoV-vaccines-SAGE-recommendation-AZD1222-2021.1-eng.pdf?sequence=5&isAllowed=y (accessed on 30 August 2021).
35. Voysey, M.; Clemens, S.A.C.; Madhi, S.A.; Weckx, L.Y.; Folegatti, P.M.; Aley, P.K.; Angus, B.; Baillie, V.L.; Barnabas, S.L.; Bhorat, Q.E.; et al. Single-dose administration and the influence of the timing of the booster dose on immunogenicity and efficacy of ChAdOx1 nCoV-19 (AZD1222) vaccine: A pooled analysis of four randomised trials. *Lancet* **2021**, *397*, 881–891. [CrossRef]
36. Government of Canada. Archived 10: Extended Dose Intervals for COVID-19 Vaccines to Optimize Early Vaccine Rollout and Population Protection in Canada in the Context of Limited Vaccine Supply. Available online: https://www.canada.ca/en/public-health/services/immunization/national-advisory-committee-on-immunization-naci/extended-dose-intervals-covid-19-vaccines-early-rollout-population-protection.html (accessed on 30 August 2021).
37. COVID-19 Vaccination Programme: FAQs on Second Doses. Available online: https://www.england.nhs.uk/coronavirus/wp-content/uploads/sites/52/2021/03/C1254-covid-19-vaccination-programme-faqs-on-second-dose-v2.pdf (accessed on 23 August 2021).
38. Wang, R.; Chen, J.; Gao, K.; Wei, G.-W. Vaccine-Escape and Fast-Growing Mutations in the United Kingdom, the United States, Singapore, Spain, India, and Other COVID-19-Devastated Countries. *Genomics* **2021**, *113*, 2158–2170. [CrossRef] [PubMed]

Article

Age, Disease Severity and Ethnicity Influence Humoral Responses in a Multi-Ethnic COVID-19 Cohort

Muneerah Smith [1,†], Houari B. Abdesselem [2,3,†], Michelle Mullins [1,†], Ti-Myen Tan [4,†], Andrew J. M. Nel [1], Maryam A. Y. Al-Nesf [5], Ilham Bensmail [2], Nour K. Majbour [2], Nishant N. Vaikath [2], Adviti Naik [2], Khalid Ouararhni [2], Vidya Mohamed-Ali [6], Mohammed Al-Maadheed [6], Darien T. Schell [1], Seanantha S. Baros-Steyl [1], Nur D. Anuar [4], Nur H. Ismail [4], Priscilla E. Morris [4], Raja N. R. Mamat [4], Nurul S. M. Rosli [4], Arif Anwar [4], Kavithambigai Ellan [7], Rozainanee M. Zain [7], Wendy A. Burgers [8,9,10], Elizabeth S. Mayne [11], Omar M. A. El-Agnaf [2,*] and Jonathan M. Blackburn [1,4,10,*]

1 Department of Integrative Biomedical Sciences, Faculty of Health Sciences, University of Cape Town, Cape Town 7925, South Africa; muneerah.smith@uct.ac.za (M.S.); MLLMIC052@myuct.ac.za (M.M.); andrew.nel@uct.ac.za (A.J.M.N.); SCHDAR006@myuct.ac.za (D.T.S.); BRSSEA001@myuct.ac.za (S.S.B.-S.)

2 Neurological Disorders Research Center, Qatar Biomedical Research Institute (QBRI), Hamad Bin Khalifa University, Qatar. Foundation, Doha P.O. Box 34110, Qatar; habdesselem@hbku.edu.qa (H.B.A.); ibensmail@hbku.edu.qa (I.B.); nmajbour@hbku.edu.qa (N.K.M.); nvaikath@hbku.edu.qa (N.N.V.); anaikjana@hbku.edu.qa (A.N.); kouararhni@hbku.edu.qa (K.O.)

3 Proteomics Core Facility, Qatar Biomedical Research Institute (QBRI), Hamad Bin Khalifa University, Qatar Foundation, Doha P.O. Box 34110, Qatar

4 Sengenics Corporation, Level M, Plaza Zurich, Damansara Heights, Kuala Lumpur 50490, Malaysia; t.myen@sengenics.com (T.-M.T.); n.diana@sengenics.com (N.D.A.); science@sengenics.com (N.H.I.); Priscilla@sengenics.com (P.E.M.); r.shirin@sengenics.com (R.N.R.M.); shiela@sengenics.com (N.S.M.R.); a.anwar@sengenics.com (A.A.)

5 Hamad General Hospital, Hamad Medical Corporation, Doha P.O. Box 3050, Qatar; Mariamali@hamad.qa

6 Anti-Doping Laboratory Qatar, Sports City Road, Aspire Zone, Doha P.O. Box 27775, Qatar; VALI@adlqatar.qa (V.M.-A.); Mohammed.AlMaadheed@adlqatar.qa (M.A.-M.)

7 Virology Lab, Level 2, Block C7, Infectious Disease Research Centre, Institute for Medical Research, Setia Alam, Selangor 40170, Malaysia; kavithambigai@moh.gov.my (K.E.); rozainanee@moh.gov.my (R.M.Z.)

8 Division of Medical Virology, Department of Pathology, University of Cape Town, Cape Town 7925, South Africa; wendy.burgers@uct.ac.za

9 Wellcome Centre for Infectious Diseases Research in Africa, University of Cape Town, Cape Town 7925, South Africa

10 Institute of Infectious Disease and Molecular Medicine, Faculty of Health Sciences, University of Cape Town, Cape Town 7925, South Africa

11 Department of Immunology, National Health Laboratory Service (NHLS) and University of the Witwatersrand, Charlotte Maxeke Johannesburg Academic Hospital, Johannesburg 2196, South Africa; elizabeth.mayne@nhls.ac.za

* Correspondence: oelagnaf@hbku.edu.qa (O.M.A.E.-A.); jonathan.blackburn@uct.ac.za (J.M.B.); Tel.: +97-455-935-568 (O.M.A.E.-A.); +27-214-066-071 (J.M.B.)

† These authors contributed equally to the work.

Citation: Smith, M.; Abdesselem, H.B.; Mullins, M.; Tan, T.-M.; Nel, A.J.M.; Al-Nesf, M.A.Y.; Bensmail, I.; Majbour, N.K.; Vaikath, N.N.; Naik, A.; et al. Age, Disease Severity and Ethnicity Influence Humoral Responses in a Multi-Ethnic COVID-19 Cohort. *Viruses* **2021**, *13*, 786. https://doi.org/10.3390/v13050786

Academic Editors: Burtram C. Fielding and Georgia Schäfer

Received: 25 March 2021
Accepted: 26 April 2021
Published: 28 April 2021

Abstract: The COVID-19 pandemic has affected all individuals across the globe in some way. Despite large numbers of reported seroprevalence studies, there remains a limited understanding of how the magnitude and epitope utilization of the humoral immune response to SARS-CoV-2 viral anti-gens varies within populations following natural infection. Here, we designed a quantitative, multi-epitope protein microarray comprising various nucleocapsid protein structural motifs, including two structural domains and three intrinsically disordered regions. Quantitative data from the microarray provided complete differentiation between cases and pre-pandemic controls (100% sensitivity and specificity) in a case-control cohort (n = 100). We then assessed the influence of disease severity, age, and ethnicity on the strength and breadth of the humoral response in a multi-ethnic cohort (n = 138). As expected, patients with severe disease showed significantly higher antibody titers and interestingly also had significantly broader epitope coverage. A significant increase in antibody titer and epitope coverage was observed with increasing age, in both mild and severe disease, which is promising for vaccine efficacy in older individuals. Additionally, we observed significant differences in the breadth and strength of the humoral immune response in relation to ethnicity, which may reflect differences in genetic and lifestyle factors. Furthermore, our data enabled localization of the immuno-dominant epitope to the C-terminal structural domain of the viral nucleocapsid protein in two independent cohorts. Overall, we have designed, validated, and tested an advanced serological assay that enables accurate quantitation of the humoral response post natural infection and that has revealed unexpected differences in the magnitude and epitope utilization within a population.

Keywords: immunoassay; SARS-CoV-2 nucleocapsid protein; epitope coverage; quantitative antibody binding; protein microarray; SARS-CoV-2 antibodies; humoral response

1. Introduction

On the 30 January 2020, a public health emergency was declared by the World Health Organization (WHO) following extensive laboratory tests that led to the identification of a novel coronavirus, SARS-CoV-2, as the causative agent of pneumonia in Wuhan, China [1]. The virus can be spread from person-to-person via direct transmission of respiratory droplets or indirectly via contact with contaminated surfaces [2]. A global pandemic was declared in March 2020, leading to extreme measures to control the spread of coronavirus disease 2019 (COVID-19) [3], which in turn has had a negative effect on global economies, medical infrastructures, and mental health [4]. This has increased the need to understand the kinetics of the immune response to COVID-19. As of 12 March 2021, the coronavirus has spread to 221 countries and territories, affecting 119,165,187 people globally, and has been the cause of approximately 2,642,905 deaths [5].

Certain comorbidities have been associated with more severe COVID-19 symptoms and worse disease prognosis; therefore, understanding the underlying mechanisms for disease progression, including innate and adaptive immune responses, is of utmost importance to protect vulnerable individuals [6,7]. Furthermore, both differences in gender and ethnicity may influence disease susceptibility and mortality [8]. Classically, antigen-specific T-cells are considered the first line of adaptive responses to a new viral infection and act to limit disease severity and control disease progression, with antigen-specific $CD8^+$ T-cells able to target and kill virally infected host cells; direct T-cell killing of viral particles is however less common. By contrast, the proliferation of antigen-specific B-cells takes longer, since it requires help from cognate $CD4^+$ T-cells, but results ultimately in the secretion of high-affinity antigen-specific antibodies that can directly opsonize viral particles in peripheral fluids and mucosal tissues, thereby targeting the virus for neutralization and/or eradication, as well as providing the basis for mucosal immunity against subsequent re-infection. B- and T-cell responses thus work in parallel and are likely equally important in primary SARS-CoV-2 infections. Interestingly, recent data from the UK COVIDsortium suggest that while most COVID-19 cases develop either neutralizing antibody or T-cell responses, the correlation between the magnitude of these responses is discordant [9]. This suggests that a more detailed understanding of both B- and T-cell responses in COVID-19 disease, as well as in subsequent immunity against re-infection by SARS-CoV-2, is still required.

In general, antigen-specific antibodies are expected to vary in titer between virally infected individuals and also to vary in target epitope and functionality—including neutralization activity (by blockade of viral-host receptor interactions), directing phagocytosis or complement-dependent killing, or agglutination. Following the COVID-19 outbreak, many antibody tests have been developed to determine the extent of current and previous SARS-CoV-2 virus infections in a given population. However, most of these antibody tests are qualitative or semi-quantitative mono-epitope tests and are unable to localize antibody binding or characterize the breadth of epitope coverage in individual patients. Given the current global interest in the age-dependence and durability of humoral responses to natural infection and to vaccination, there therefore remains a need for new, advanced serology assay platforms that can assist in quantifying the complexity of the antibody responses to COVID-19 disease.

Screening for immunoreactivity utilizing a high-throughput antigen microarray in principle enables the simultaneous assay of multiple discrete, folded domains and epitopes of a given antigen, thus potentially allowing identification of antibody correlates of ongoing protection and of development of durable immunity against subsequent SARS-CoV-2 infection. Furthermore, using pre-pandemic and known negative samples, it is possible to identify sources of cross-reactivity, which can be utilized to re-engineer functional epitopes to decrease the rate of false positives; however, this risks decreasing the sensitivity by the removal of true target epitopes. Recent studies utilizing various protein array platforms have reported high specificity and sensitivity [10–12]; however, these previous platforms lack the ability to quantitate differential antibody epitope utilization—including both linear and discontinuous epitopes—across cohorts of convalescent COVID-19 patients.

In addition, due to the high sequence similarity between SARS-CoV-1 and SARS-CoV-2 [13], there is a potential for antibody cross-reactivity between SARS-CoV-1 antibodies and SARS-CoV-2 antigens in regions where the original SARS outbreak was prevalent. However, a previous study reported that SARS-CoV-1 specific antibodies were undetectable in 91% of samples tested six years following infection [14]. Furthermore, there were a total of only 8096 SARS-CoV-1 cases worldwide, and SARS-CoV-1 has not circulated in the human population for over 17 years [15]; therefore, the chances of false positives in serological assays due to cross-reactivity are very low. In contrast, the seroprevalence of antibodies against naturally circulating human coronaviruses (hCoVs) is ubiquitous in most individuals [16], making the possible immune cross-reactivity between the four common hCoVs (229E, NL63, OC43, and HKU1), SARS-CoV-1, MERS, and SARS-CoV-2 an important factor in the design of immunoassays.

Here, we have designed and validated a novel, quantitative, sensitive, and specific SARS-CoV-2 multi-epitope fluorescent immunoassay, based on the nucleocapsid protein. The array is based on the use of the biotin carboxyl carrier protein (BCCP), which acts as a marker for the correct folding of proteins, since only correctly folded proteins will be biotinylated. Therefore, it is possible to control the immobilization of antigens onto a streptavidin coated surface in an oriented manner [17]. Different prototype array designs, using various engineered SARS-CoV-2 nucleocapsid protein structural motifs, were tested on a cross-sectional convalescent COVID-19 cohort and pre-pandemic controls to determine cross-reactivity. The specificity and sensitivity of the final array design were validated in an independent cohort. We then used this SARS-CoV-2 antigen microarray platform to explore the relationship between clinical data—age, disease severity, and ethnicity—and quantitative, epitope-specific antibody titers in a cohort of COVID-19 patients drawn from a migrant worker population in a single geographic region.

2. Materials and Methods

2.1. Study Design

Three different COVID-19 cohorts were used to develop, validate, and utilise the immunoassay.

2.1.1. Cohort 1

Serum or plasma were prepared from blood samples collected from a cross-sectional cohort of 106 convalescent COVID-19 patients, recruited from Gauteng and Western Cape, South Africa, and stored at –80 °C until further analysis. The clinical characteristics of this cohort are summarized in Table 1. These patients were originally tested for SARS-CoV-2 using reverse transcriptase polymerase chain reaction (RT-PCR), using upper respiratory tract samples (nose or throat). These serum/plasma samples were used to design and develop the prototype array platform. Ethical approvals for these studies were obtained from the Human Research Ethics Committees of the University of Witwatersrand (M200468) and the University of Cape Town (UCT; HREC 210/2020). All patients provided written, informed consent. The plasma of 58 pre-pandemic colorectal cancer (CRC) patients and 10 healthy volunteers were used as additional controls for developing the array platform (UCT ethics approval HREC 269/2011).

Table 1. Clinical characteristic of COVID-19 patient cohorts.

Clinical Characteristics		Cohort 1	Cohort 2	Cohort 3
Total number of patients		**174**	**100**	**138**
Disease status	Pre-pandemic disease controls	68	50	0
	COVID-19 PCR − ve	23		
	COVID-19 PCR + ve	76	50	100
	No COVID-19 PCR test data	7		38
Disease Severity	Asymptomatic (PCR − ve)	4	0	0
	Symptomatic (PCR − ve)	19	0	0
	Asymptomatic (PCR + ve)	14	0	7
	Mild (PCR + ve)	24	0	43
	Severe (PCR + ve)	34	50	50
	Asymptomatic (no PCR test data)	7	0	38
	Not declared (PCR + ve)	4	0	0
Gender	Female	55 *	30 *	12
	Male	49 *	13 *	126
	Not declared	2 *	7 *	0
Age distribution	18–40	60 *	10 *	67
	41–60	38 *	24 *	65
	61–73	6 *	9 *	6
	Not declared	2 *	7 *	0
Ethnicity	African	9 *	0	
	Caucasian	72 *	0	0
	Colored	1 *	0	0
	Half-Japanese, half-Caucasian	1 *	0	0
	South Asian	9 *	100	94
	Middle East (Other)	0 *	0	10
	Middle East (Qatari)	0 *	0	18
	Other	0 *	0	15
	Not declared	14 *	0	1

* Convalescent PCR positive patients.

2.1.2. Cohort 2

The validation study was performed using sera collected from fifty randomly selected, hospitalized, PCR-positive COVID-19 patients with severe disease as part of the standard of care at Hospital Sungai Buloh, Selangor, Malaysia. The clinical characteristics of the patients in the cohort are summarized in Table 1. Fifty pre-pandemic HIV positive serum samples were used as true negative controls. In this cohort, no additional clinical annotations were provided.

2.1.3. Cohort 3

Hospitalized COVID-19 positive patients (n = 100) admitted to Hamad Medical Corporation hospitals in Doha, Qatar, with confirmed positive RT-PCR results (sputum and throat swab) for the SARS-CoV-2 virus were randomly selected and enrolled for this study. The demographics of this cohort were therefore expected to be representative of COVID-19 cases in Qatar and included individuals from various ethnic groups (Middle Eastern (Qatari), Middle Eastern (non-Qatari), South Asian, and other). Peripheral blood was collected within five to seven days of admission and processed into plasma and serum, and then stored at −80 °C until further analysis. Patients were classified as having either mild/moderate disease (n = 50) or severe disease (admitted to intensive care unit; n = 50). Four patients were deceased from the severe group. Blood samples from age, gender, and ethnicity matched healthy volunteers (*n* = 38) with no prior COVID-19 infection history and with normal oxygen saturation and vital signs were recruited by the Anti-Doping Laboratory Qatar (ADL-Q) for blood collection. Individuals with medical history or with cognitive disability were excluded. The clinical characteristics of COVID-19 and healthy participants are summarized in Table 1.

All participants (patients and controls) provided written informed consent prior to enrolment in the study. Ethical approval for these studies was obtained from the Hamad Medical Corporation Institutional Review Board Research Ethics Committee (reference MRC-05-003).

2.2. Selection, Cloning, and Expression of SARS-CoV-2 Antigens

2.2.1. Antigen Selection for Immunoassay Platform

Full-length SARS-CoV-2 nucleocapsid protein (UniProt accession number P0DTC9), as well as the core structural domains of the N protein (annotated as N-core) (44–362 aa), the N- terminal domain (NTD) (43–179 aa), the C-terminal domain (CTD) (246–363 aa), and 17 tiling peptides consisting of predicted B-cell epitopes in the intrinsically disordered regions (IDRs; including peptides spanning residues 395–412, 211–228, and 367–389) were selected for inclusion on the prototype array design.

2.2.2. Gene Synthesis and Cloning

The full-length SARS-CoV-2 nucleocapsid (N) gene was synthesized (GeneArt, Regensburg, Germany) and cloned into a proprietary *Escherichia coli/ Spodoptera frugiperda* transfer vector, pPRO8, such that the construct encoded the full-length N protein as an in-frame fusion to a C-terminal Biotin Carboxyl Carrier Protein (BCCP) and c-Myc tag. pPRO8 is a derivative of pTriEx1.1 (Sigma, St Louis, MO, USA) and encodes the *E. coli* BCCP domain (amino acids 74–156 of the *E. coli accB* gene) downstream of a viral polyhedrin promoter and cloning sites; flanking this *polh*-BCCP expression cassette are the baculoviral 603 gene and the 1629 genes to enable subsequent homologous recombination of the construct into a replication-deficient baculoviral genome [17].

N-core, NTD, and CTD clones were constructed from the full-length N gene using the oligo pairs summarized in Table S1. Amplicons were generated by polymerase chain reaction using Vent DNA polymerase (New England Biolabs, Ipswich, MA, USA), digested with *Spe*I and *Nco*I (New England Biolabs) restriction enzymes and ligated into the equivalent sites in pPRO8, using standard protocols. All generated clones thus encoded N-protein structural motifs as in-frame fusions to a C-terminal BCCP c-Myc tag. In addition, seventeen tiling peptides ('IDRs 1 to 17') were synthesized with an N-terminal biotin moiety (Synpeptide, Shanghai, China) (Table S2).

2.2.3. Expression of Nucleocapsid Proteins as Fusions to a BCCP Tag

Following co-transfection of *S. frugiperda* Sf9 cells with a relevant pPRO8-derived transfer vector plus a linearized, replication deficient bacmid vector (*Autographa californica* baculovirus vector pBAC10:KO_{1629} [17]), baculovirus was amplified and recombinant proteins were expressed in *S. frugiperda* superSf9–3 strain (Oxford Expression Technologies, Oxford, UK) using previously published protocols [17]. Clarified cell lysates were prepared in insect lysis buffer (25 mM Hepes, 50 mM KCL, 20% glycerol, 0.1% Triton × 100, 1 × Halt™ Protease Inhibitor Cocktail, EDTA-free (Thermo Scientific, Waltham, MA, USA), 0.25% sodium deoxycholate acid, 25 U/mL Pierce Universal nuclease (Thermo Scientific), pH 8). Expression yields and *in vivo* biotinylation of each antigen were assessed by Western blot using a streptavidin-HRP conjugate probe (GE Healthcare, Chicago, IL, USA) (Figure S1). Lysates were stored at −80 °C before array printing. Peptides were solubilized in the same buffer (without nuclease and protease inhibitor) at a final concentration of 0.1 mg/mL. Control antigens used in the microarray included 50 µg/mL of biotinylated human immunoglobulins G, A, and M (hIgG, hIgA, and hIgM, respectively; Rockland, Gilbertsville, PA, USA) and 132 µg/mL of biotinylated anti-human immunoglobulin G (anti-hIgG; Rockland) as well as in house derivatized NHS-ester-Cy3 (Thermo Scientific) biotinylated BSA (Cy3-BSA) at 40 μg/mL.

2.3. Fabrication of Prototype and Final Protein Microarray

Prototype microarrays were printed using a QArray2 printer (Molecular Devices, San Jose, CA, USA) using methods described previously [18] on proprietary streptavidin-coated hydrogel slides (7.5 × 2.5 cm; Sengenics Corporation, Singapore). Each antigen was printed in triplicate with a mean size of 450 μm per spot. Eight replica arrays were printed per slide. After printing, the slides were incubated in a blocking buffer (20% Glycerol, 25 mM HEPES buffer (pH 7.4), 50 mM KCl, 1% Triton X-100, 1 mM DTT and 50 μM Biotin) and stored at 4 °C until used.

The final array layout (Figure S2) was fabricated using piezo-electric printing technology (Biodot, Irvine, CA, USA) onto streptavidin-coated hydrogel slides. Each antigen was printed in triplicate in a 24-plex format (i.e., 24 replica arrays per slide) with a mean size of 125 μm per spot. Slides were blocked and stored at −20 °C in blocking buffer (25 mM HEPES, 50 mM KCl, 4 mM $CaCl_2$, 20 mM $MgCl_2$, 20% Glycerol, 0.2% Triton X-100, 2% BSA). Successful immobilization and in situ purification of biotinylated proteins from lysates were confirmed via an anti-c-Myc (Sigma) assay.

2.4. Serological Assays

Optimization of Serum Concentration and Determination of Linear Range

For serial dilution assays, the serum or plasma was diluted 1:50, 1:100, 1:200, or 1:400 before adding it to the slides and commencing with the hybridization assay, as described below. All prototype microarrays were developed measuring IgG responses using 20 μg/mL AlexaFluor (AF) 647-labeled anti-human IgG. Notably, we observe no significant difference in performance of our immunofluorescence assays with serum or plasma (data not shown) and consider the assay to be equally compatible with both.

Microarray slides were washed with PBST (PBS, 0.2% Tween-20, pH 7.4) at RT for 3 × 5 min with gentle agitation, then dried by centrifugation at 1200× *g* for 2 min. Individual arrays were isolated using ProPlate 24 plex multi-well chambers (GraceBio-Labs, Bend, OR, USA). Prior to assays, serum samples were incubated with 0.1% Triton X-100 for 1 h on ice to deactivate potential live virions, then diluted 1:50 in assay buffer (PBST, 0.1% BSA, 0.1% milk powder). Individual arrays were incubated with 50 μL diluted serum for 1 h at RT with gentle agitation, then briefly rinsed with PBST, after which the slides were removed from the gaskets, washed for 3 × 5 min in PBST and dried by centrifugation at 1200× *g* for 2 min.

Arrays were then incubated with detection antibody (20 μg/mL Cy3-labeled anti-human IgG in assay buffer) for 30 min at RT with gentle agitation. The wells were briefly rinsed with PBST, after which the slides were removed from the gaskets and washed for 3 × 5 min in PBST with gentle agitation and dried by centrifugation at 1200× *g* for 2 min.

2.5. Bioinformatic Analysis

2.5.1. Image Analysis: Raw Data Extraction

Slides were scanned at a fixed gain setting using either an InnoScan 710 (Innopsys, Carbonne, France) or G2505C (Agilent, Santa Clara, CA, USA) fluorescence microarray scanner, generating a 16-bit TIFF file. A visual quality control check was conducted, and any arrays showing spot merging or other artefacts were re-assayed.

A GAL (GenePix Array List) file containing information regarding the location and identity of all probed spots was used to aid with image analysis. Automatic extraction and quantification of each spot were performed using either Mapix software (Innopsys) or GenePix Pro 7 (Molecular Devices) software, yielding the median foreground and local background pixel intensities for each spot.

2.5.2. Data Pre-Processing

The mean net fluorescence intensity of each spot was calculated as the difference between the raw mean intensity and its local background. Extrapolated data were filtered and normalized using an in-house developed software (CT100+ programme). CVs for biotinylated Cy3-BSA were routinely below 5%. Human IgG (detected by fluorescently labeled secondary antibody) and human anti-IgG (detected only when plasma or serum is added to the slide) were used as positive controls to assess image signal intensity. Thresholds for positive signals for each antigen were determined using the OptimalCutpoints package with an emphasis on maximizing specificity [19].

Reciprocal titers per-antigen were determined from measured net fluorescence intensity, based on the projected further dilution of the sample required to reach the limit of detection in the assay, according to the following equation:

$$\textit{Reciprocal Titer} = (\textit{Net Intensity (RFU)} \times \textit{initial serum dilution/limit of detection (RFU)}) \quad (1)$$

Underlying assumptions include: linearity of antibody binding signal vs. serum dilution, as observed both in this work and previously on protein arrays with the same underlying architecture [20]; linearity of signal observed for the dilution series of biotinylated hIgG controls on protein arrays with the same underlying architecture, in accordance with ligand binding theory (data not shown); and an assumed limit of detection of 50 RFU (equating to the noise threshold of the surrounding background). A cumulative score was then calculated based on the sum of reciprocal titers for non-overlapping domains of the N antigens to determine the seropositivity of a given sample.

2.5.3. Statistical Tests

Sensitivity, specificity, and confidence intervals estimate were estimated using previously reported methodologies [21]. Other statistical analyses and graphical representation were generated using the R programming language (v 4.0.2) and GraphPad Prism (v 9.0; GraphPad Software, San Diego, CA, USA). Pearson's correlation was performed to establish correlations between cumulative titer and various variables. Either the Wilcoxon–Mann–Whitney test or a one-way ANOVA with Welches correction was applied to determine the statistical significance of the differences observed between multiple independent groups (HC, mild and severe or case vs. control).

3. Results

3.1. Developing a High-Sensitivity, High-Specificity SARS-CoV-2 Antigen Microarray

It has previously been estimated that roughly 90% of B-cell epitopes are discontinuous [22,23] and surface exposed, yet it is well known that antibodies have a propensity for binding non-specifically to normally buried hydrophobic surfaces that become exposed on unfolded proteins. In order to allow for antibody recognition of discontinuous as well as linear surface exposed epitope, while minimizing non-specific binding, we fused full-length and functional domains of the SARS-CoV-2 nucleocapsid protein to a C-terminal Biotin Carboxyl Carrier Protein (BCCP) tag and expressed the resultant fusion proteins in insect cells. BCCP is only biotinylated *in vivo* when correctly folded [24], and misfolded fusion proteins have been shown to result in misfolding of BCCP; thus, only correctly folded fusion proteins become biotinylated and bind to a streptavidin-coated surface [17].

3.1.1. Selecting N-Protein Constructs for the Final Microarray Design

The IgG response to SARS-CoV-2 full-length N protein was compared between pre-pandemic healthy controls (HC) and convalescent COVID-19 patients (P) drawn from Cohort 1. A serial dilution (1:50, 1:100, 1:200, 1:400) of pooled samples from the 10 HC and 10 P samples was performed to assess overall signals (Figure S3A). Although the signal is higher for the Ps than the HCs, high relative fluorescent units (RFU) signals were detected for both sample sets, which was confirmed for the individual HC and P samples as shown in Figure S3B.

An additional three SARS-CoV-2 N-protein constructs were therefore cloned, expressed, and purified/immobilized on the microarray, corresponding to the core structural domains ('N-core'; residues 44–362), as well as the isolated N-terminal domain (residues 43–179) and C-terminal domains (residues 246–363; Figure S2). Domain boundaries in the SARS-CoV-2 nucleocapsid protein were identified by ClustalW-based sequence alignment of the SARS-CoV-1 (UniProt ID: P59595) and SARS-CoV-2 (UniProt ID: P0DTC9) nucleocapsid protein sequences and comparison with published structures of the SARS-CoV-1 nucleocapsid protein (PDB IDs: NTD, 1SSK; CTD, 2CJR).

We determined the optimal serum concentration for antibody binding to these new antigens using a serum dilution series from 1:50 to 1:12800. Figure S4 shows representative ligand (i.e., antibody) binding curves for two randomly selected samples from Cohort 1 (P189 and P192). For P189, the highest dilution that still gave signal above background for the three N-protein constructs was 1:6400 dilution, with signal beginning to saturate at 1:100 dilution (Figure S4A). For P192, the highest dilution that still gave signal above background was 1:400, and signal was still in the linear range at 1:50 dilution (Figure S4B). We used 1:50 serum dilution for all subsequent assays.

These additional protein constructs also allowed us to assess non-specific binding and epitope coverage. Here, selected plasma samples from eight colorectal cancer patients (Cohort 1) were used as disease controls (C) and compared to seven Ps (Figure S5). The RFU signals for Cs were similar, ranging from 786–3855 and 639–3376 RFU for the full-length N protein (no PLS) and truncated N protein, respectively. However, the RFU signal for Ps was higher for the truncated N protein (3615–36993 RFU) compared to the full-length N protein (3034–12405), suggesting that the truncated N protein could offer a similar level of specificity, but a higher level of sensitivity compared to the full-length N protein. The C- and N-terminal domains display lower levels of non-specific binding with RFU levels ranging from 154–1050 and 219–1684 RFU for the Cs, respectively. However, the RFU signal for the Ps also decreased, ranging from 1011–16845 and 560–5161 for the C- and N-terminal domains, respectively.

3.1.2. Selecting Peptides from the N Protein for Microarray Fabrication

To further improve the sensitivity and specificity of the platform, and to determine epitope coverage, a microarray was fabricated with 17 biotinylated peptides (Table S2) derived from the N protein, which were predicted B-cell epitopes [25]. The IgG response to these 17 peptides was initially assessed using 10 HCs and 15 Ps (Figures S6–S22). Varying degrees of non-specific binding were observed for 14 of the peptides, whereas Peptides 2, 6, and 8 showed little or no non-specific binding for the HCs, and a linear response with serum dilution for Ps. Two peptides (Peptides 5 and 10, both of which are lysine- and arginine-rich and have strongly basic patches) were observed to bind non-specifically and with high titers to pre-pandemic disease control sera, as well as to anti-human IgG, anti-His, and anti-c-myc antibodies: these two peptides flank the core structural domains of the nucleocapsid protein and may thus explain the significant cross-reactivity of the full-length SARS-CoV-2 N protein observed here with pre-pandemic sera (Figure S3). Peptides 1, 3, and 16 showed some non-specific binding, but some Ps who were non-responsive to Peptides 2, 6, and 8 were found to be responsive to Peptides 1, 3, or 16. Thus, Peptides 1, 2, 3, 6, 8, and 16 were retained for further analysis.

To evaluate which predicted N-protein B-cell epitopes resulted in the highest frequencies of disease-specific antibody binding, samples from 91 Ps and 58 Cs were then assayed against Peptides 1, 2, 3, 6, 8, and 16 (Figure S23). Nine Ps (RFU range: 301–2885) and two Cs (RFU range: 843–2623) produced an IgG response to Peptide 1; 27 Ps (RFU range: 138–62833) and four Cs (RFU range: 165–18245) produced an IgG response to Peptide 2; 15 Ps (RFU range: 123–64465) and 11 Cs (RFU range: 122–7704) produced an IgG response to Peptide 3. Notably, the frequency of positive signals amongst the Ps to Peptides 1, 2, and 3 was relatively low, while the magnitude of the IgG signal from the majority of Ps to these peptides was also found to be low and in the same range as signal from the Cs, suggesting that these peptides were not suitable for further development. By contrast, 45 and 41 Ps, respectively, displayed a moderate to high IgG response to Peptides 6 and 8, while only four Cs displayed low IgG responses towards either (RFU range: 141–1012), indicating that these peptides individually should have a high specificity and a moderate sensitivity. Finally, although a median signal of ~2500 RFU was found with 12 Cs for peptide 16, 41 Ps produced signals > 5000 RFU, including a number of Ps that were not reactive to peptides 6 or 8, indicating that the signal from true positives was well above the non-specific binding threshold and that Peptide 16 thus provided useful incremental benefit over Peptides 6 and 8.

Serial dilution assays using samples P189 and P192 demonstrated linearity of IgG binding to Peptides 6, 8, and 16 in the range 1:400 to 1:50 (Figure S4C,D). We therefore elected to retain Peptides 6, 8, and 16 in our design, as a means to maximize the sensitivity and specificity of the final microarray platform (Figure S2).

3.2. Technical Performance of the SARS-CoV-2 Antigen Microarray Platform in an Independent Validation Cohort

The IgG cumulative titer found for the 50 severe COVID-19 cases and 50 pre-pandemic controls in Cohort 2 was used to determine the specificity and sensitivity of the arrays. Patients were defined as seropositive towards COVID-19 when the reciprocal titer for one or more N antigens were elevated above a 'Minimum Specificity = 1' threshold determined using the OptimalCutpoints package, based on the pre-pandemic control data. All 50 hospitalized COVID-19 patients were found to be seropositive, and all 50 pre-pandemic controls were found to be seronegative on the microarray platform; thus, the performance accuracy of the array was calculated to be 100% (Table 2). Figure 1 further validates the accuracy of the array, as there is a significant elevation in antibody titers to all antigenic domains in all case samples compared to the pre-pandemic controls (Figure 1A).

Table 2. Validation immunoassay data (Cohort 2). Confusion matrix showing the number of severe COVID-19 cases (n = 50) and pre-pandemic controls (n = 50) who gave a positive or negative assay result on the microarray platform, allowing calculation of clinical sensitivity and specificity.

Immunoassay Result	**COVID-19 Status**	
	Positive	**Negative**
Positive	50	0
Negative	0	50
	Sensitivity = 100%	Specificity = 100%

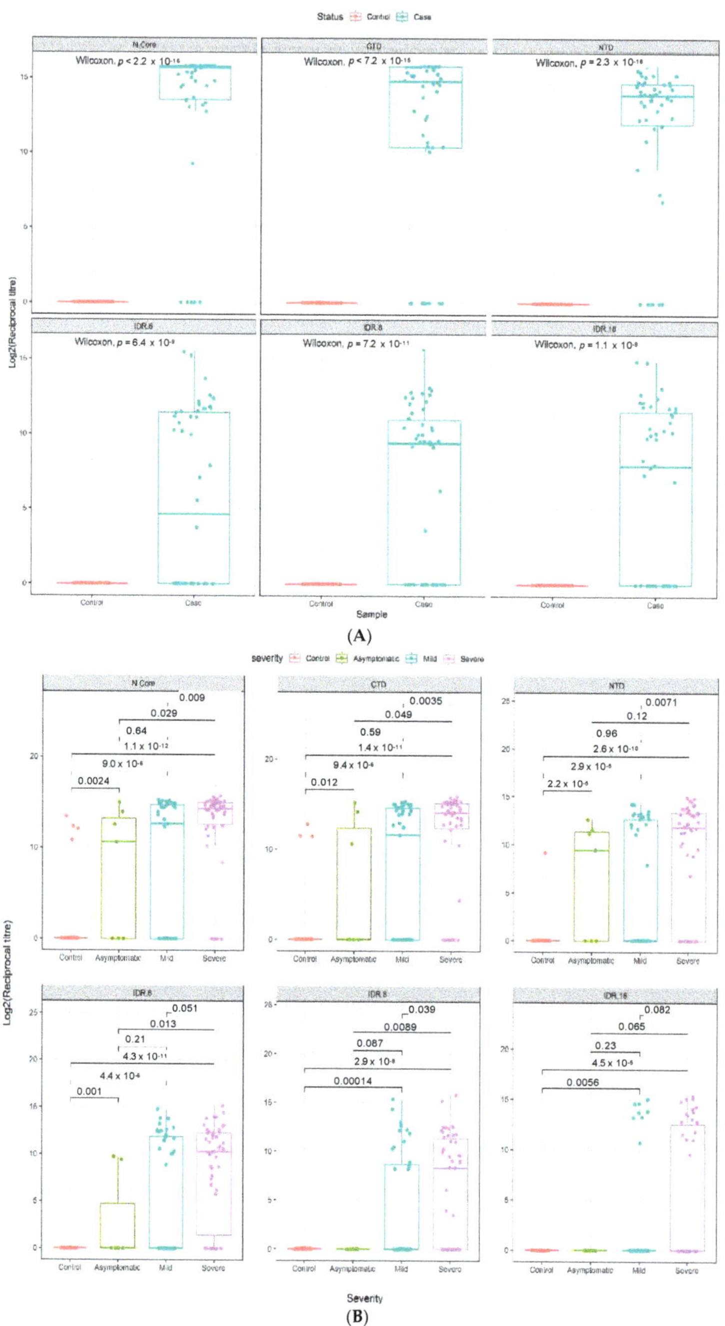

Figure 1. Epitope selectivity of IgG responses in two independent COVID-19 cohorts. Antibody reciprocal titers against different epitopes ($n = 6$) of the SARS-CoV-2 N protein in two separate COVID-19 case and control cohorts. (**A**) Validation cohort ($n = 100$), consisting of 50 hospitalized COVID-19 patients and 50 pre-pandemic controls (Cohort 2). (**B**) Multi-ethnic cohort ($n = 138$), consisting of 50 severe COVID-19 patients, 50 mild COVID-19 patients, and 38 healthy controls (Cohort 3). Boxes represent the 25th and 75th percentiles, and the midline represents the median and whiskers represent the 5th and 95th percentiles. p-values were determined using the Wilcoxon test (unpaired, two-tailed).

3.3. *Quantitative Analysis of an Independent, Multi-Ethnic Cohort Reveals Differences in Antibody Titers and Epitope Coverage Scores Associated with Age, Disease Severity, and Ethnicity*

A significant increase in antibody titers was observed between individuals with mild or severe disease and healthy controls in a further independent, multi-ethnic cohort (Cohort 3) recruited in Qatar (Figure 1B). Notably, our data reveal that the dominant antigenic epitopes lie in the two structural domains (and particularly the C-terminal domain), rather than in the intrinsically disordered regions of the nucleocapsid protein for both mild and severe disease patients in Cohorts 2 and 3, as judged by both the magnitude (reciprocal titer) and frequency of antibody recognition of the different structural motifs on our platform (Figure 1).

In Cohort 3, the nominally healthy control samples were recruited during the pandemic, rather than pre-pandemic, and were individuals with no history of COVID-19 disease but who were not tested by PCR. Four of these 38 controls were called positive by our immunoassay (Table 3), initially suggesting a specificity of 89.5%. However, closer inspection revealed that three of these four seropositive samples show significant reciprocal titers against two or more non-overlapping epitopes on the N protein (Figures 1B and 2), increasing the confidence in these controls being true positives. It therefore seems likely that these individuals in fact had prior asymptomatic SARS-CoV-2 infections, rather than representing false positive immunoassay results; the actual specificity of our immunoassay in Cohort 3 thus appears to be 97.4–100%.

Table 3. Multi-ethnic cohort immunoassay data (Cohort 3).

Disease Severity	Immunoassay Result	RT-PCR Status		Sensitivity	Specificity
		Positive	Unknown		
All samples (Case, n = 100; Control, n = 38)	Positive	75/100	4/38	0.75	0.90
	Negative	25/100	34/38		
Asymptomatic (Case, n = 7; Control, n = 38)	Positive	4/7	4/38	0.57	0.90
	Negative	3/7	34/38		
Mild (Case, n = 43; Control, n = 38)	Positive	25/43	4/38	0.58	0.90
	Negative	18/43	34/38		
Severe (Case, n = 50, Control, n = 38)	Positive	46/50	4/38	0.92	0.90
	Negative	4/50	34/38		

The sensitivity of detection found amongst PCR positive cases with mild disease (58%) or severe disease (92%; Table 3) in Cohort 3 is at first sight in line with literature expectation. However, 85% of the samples (43/50 mild; 42/50 severe) were collected within the first 14 days post onset of symptoms, and all samples were collected within 5–7 days of hospital admission. A more detailed analysis of the time to seropositivity in Cohort 3 showed a sensitivity of 75% in the first seven days post symptom onset in patients who developed severe disease, increasing to 97% by day 14 (Supplementary Table S3 and Figure S25), and a sensitivity of 56% by day 7 even in patients developing mild disease. This means that seropositivity was detected while those patients were likely still in the acute phase of infection, and we suggest that this relatively early, high sensitivity may reflect the low limits of detection achieved with our multi-epitope fluorescent immunoassay and draw attention to the high epitope coverage scores for the majority of both mild and severe seropositive patients as evidence for the basis of this technical performance (Figure 2). To further assess the performance of the assay in these five to seven day post positive PCR samples, the positive and negative predictive values were calculated and are given in Table S4.

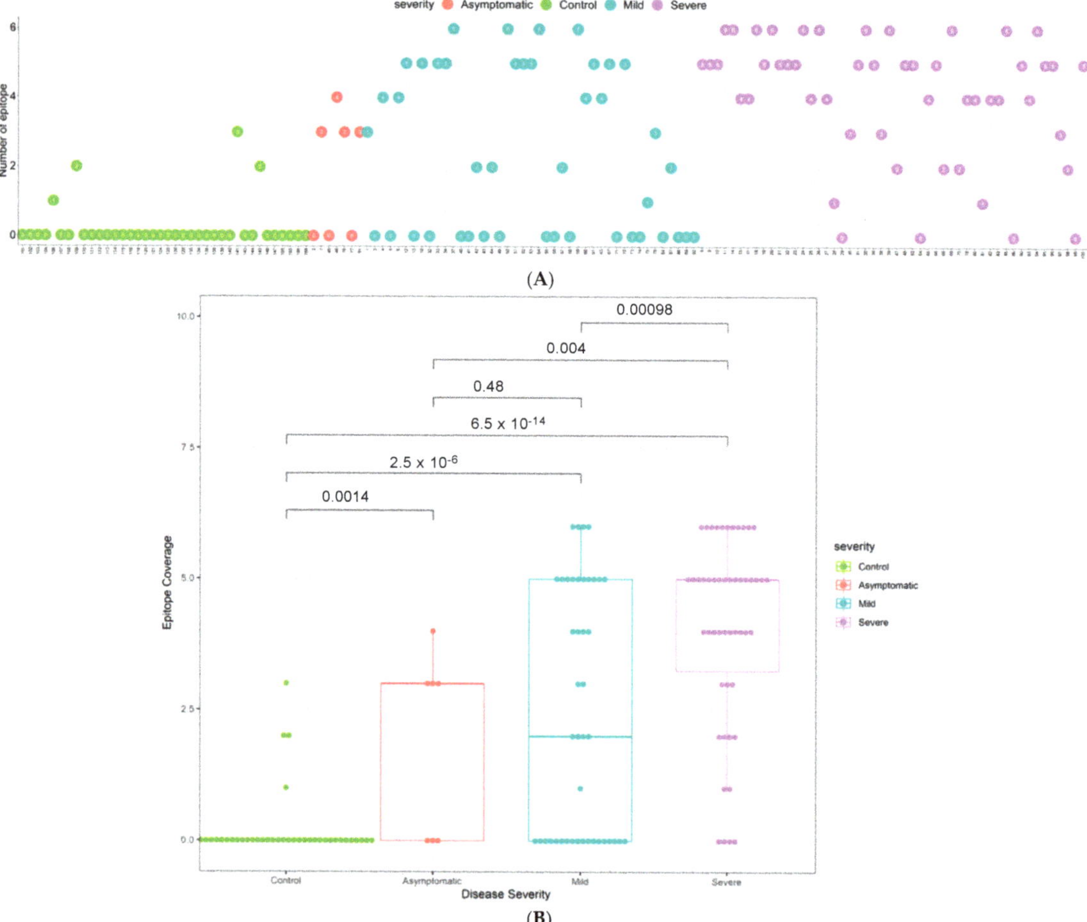

Figure 2. Epitope coverage in multi-ethnic Cohort 3. (**A**) Epitope coverage for each sample for controls, mild cases, and severe cases (n = 138). Numbers in dots represent the EPC score per participant. (**B**) Box plots displaying the epitope coverage for each disease class. p-values were determined using the Wilcoxon test (unpaired, two-tailed).

3.4. *Elevated N-Specific Antibody Titers and Broader Epitope Coverage Observed in Patients with Severe Disease*

To determine the breadth of the antibody response, the sum of the number of IgG positive epitopes was calculated for each sample and presented in Figure 2 as an Epitope Coverage (EPC) Score. Not only do patients with severe disease have significantly higher antibody titers than patients with mild disease (Figure 1B), they also respond to a broader range of epitopes (p = 0.00017; Figure 2). Furthermore, the majority of COVID-19 patients have a broader epitope coverage compared to healthy controls, and the differences in coverage are statistically significant for all comparisons (Figure 2B).

3.4.1. Increasing Antibody Titers and Epitope Coverage with Increasing Age

In both Cohorts 2 and 3, a trend to increasing antibody titer was observed with increasing age, reaching statistical significance in Cohort 2 in the age 51–60 bracket (Figure 3 and Figure S24). A similar trend was observed for the breadth of the immune response, with patients over 40, over 50, and over 60 having increasingly elevated epitope coverage scores compared to patients under 40 in Cohort 2, reaching statistical significance in the age 51–60 (p = 0.042) and >60 (p = 0.029) brackets (Figure 4A). In Cohort 3, a similar trend of increasingly elevated epitope coverage scores up to age 60 was also observed in both mild and severe disease cases (Figure 4B), but the small number of patients over 60 (n = 6) precludes robust conclusions being drawn on whether there is a genuine decline in epitope coverage scores in the >60 bracket or not.

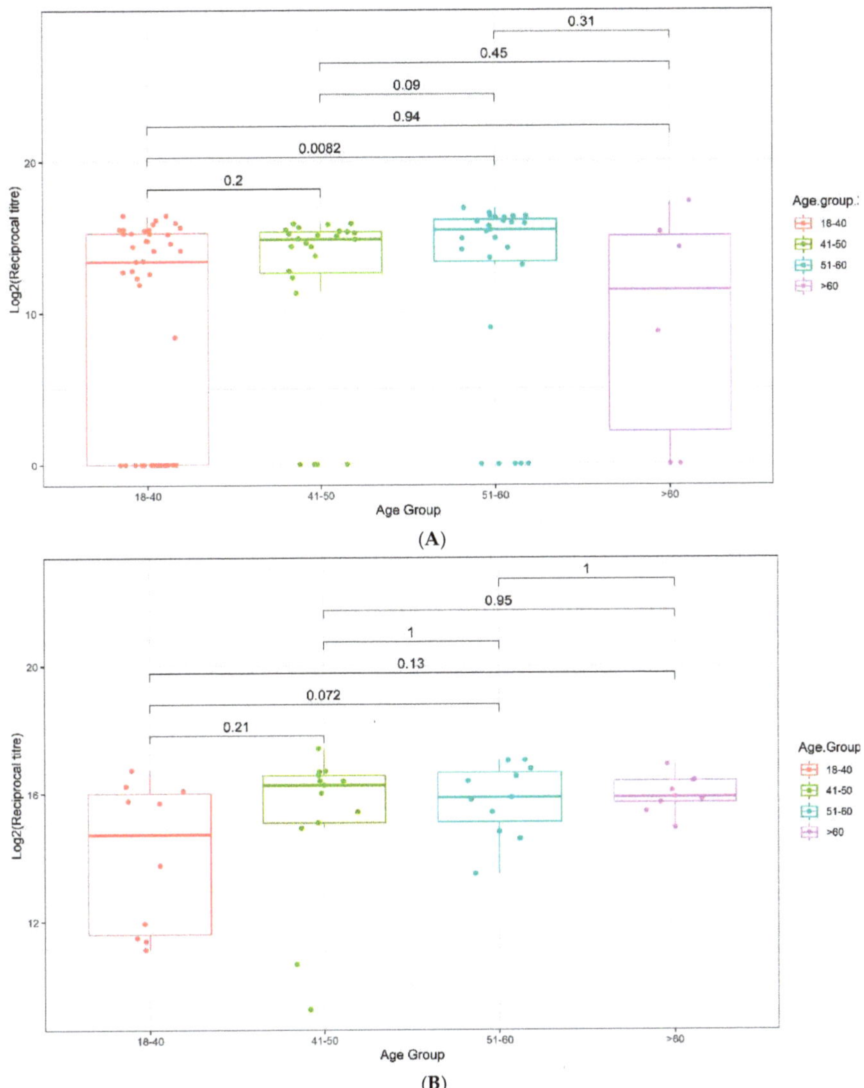

Figure 3. Box plots displaying the antibody reciprocal titers as a function of age. (**A**) Validation Cohort 2, (**B**) Multi-ethnic Cohort 3. p-values were determined using the Wilcoxon test (unpaired, two-tailed).

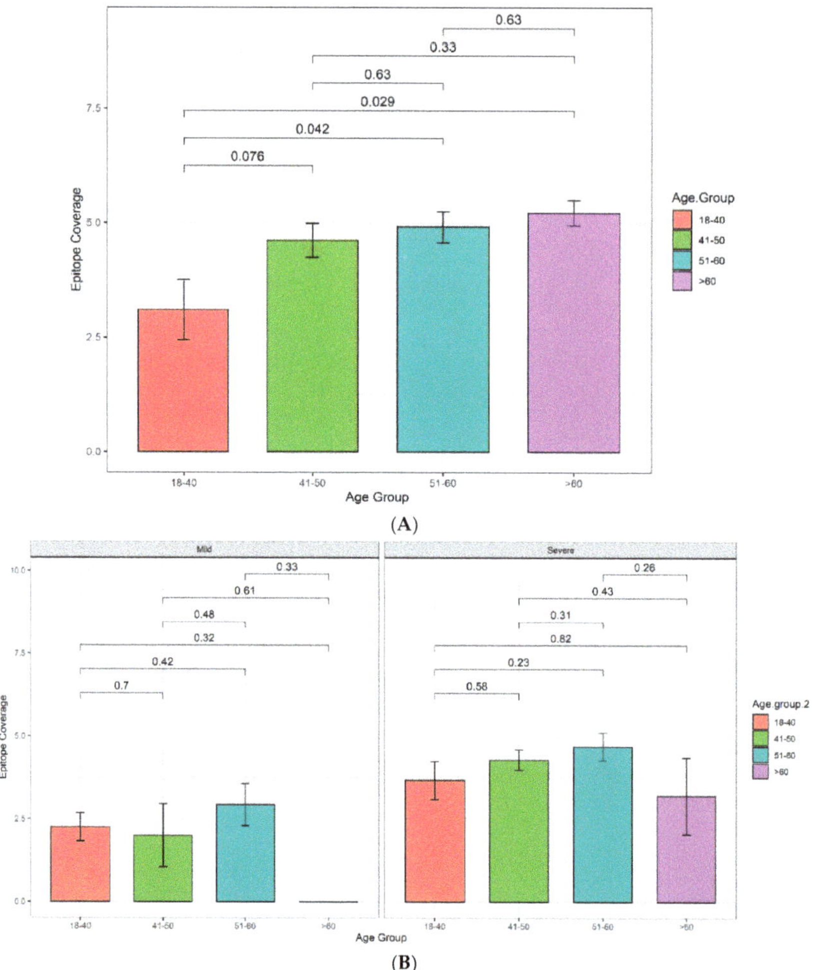

Figure 4. Histogram displaying the epitope coverage as a function of age. (**A**) Validation Cohort 2. Sample sizes: 18–40: n = 10, 41–50: n = 13, 51–60: n = 11, >60: n = 9. (**B**) Multi-ethnic Cohort 3; patients further categorized according to disease severity. Samples sizes: 18–40 mild: n = 28, severe: n= 15. 41–50 mild: n = 7, severe: n= 17. 51–60 mild: n = 14, severe: n = 13. >60 mild: n = 1, severe: n = 5.

3.4.2. The Influence of Ethnicity on N-Specific Antibody Titers and the Breadth of Epitope Coverage

The relationship between ethnicity, antibody titers, and epitope coverage was assessed, and the results are summarized in Figure 5. Of all ethnic groups assessed, the Middle Eastern ethnicity group, excluding Qatari, was the only group to display a significant increase in both antibody titers and epitope coverage in patients with severe disease in comparison to patients with mild disease (Figure 5).

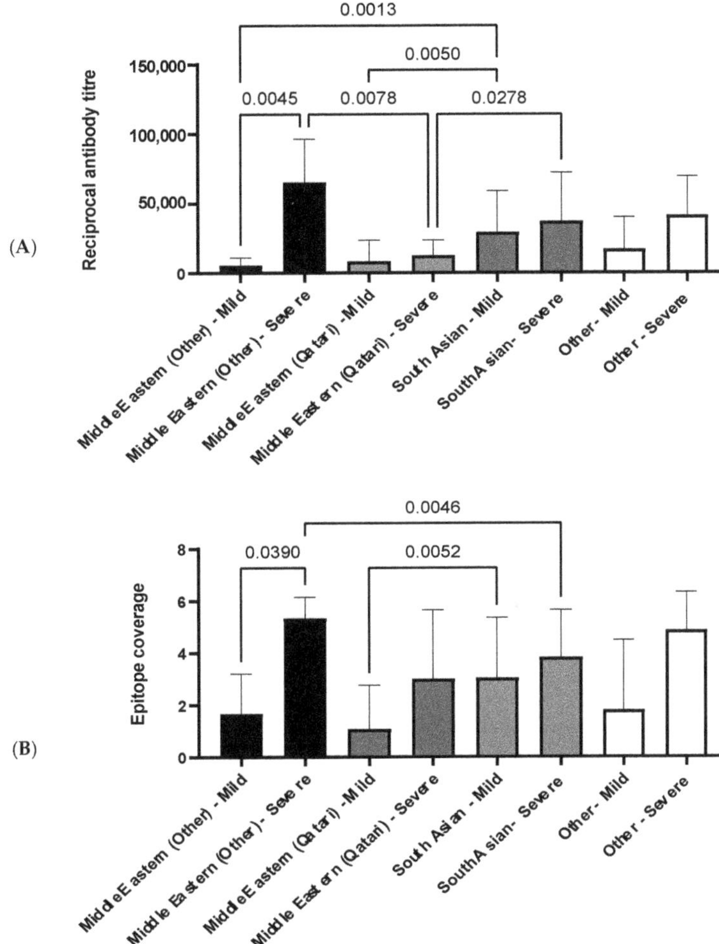

Figure 5. Histogram displaying relationship between antibody reciprocal titer and epitope coverage in Cohort 3. (**A**) Average antibody reciprocal titer for different ethnic groups and disease severities. (**B**) Average epitope coverage for different ethnic groups and disease severities. Pairwise comparisons were made using a one-way ANOVA, and p-values were calculated using Welch's correction to compare the mean of each category with each other category. Sample sizes: Middle Eastern (other) mild: $n = 2$, severe: $n = 6$. Middle Eastern (Qatari) mild: $n = 12$, severe: $n = 3$. South Asian mild: $n = 30$, severe: $n = 34$. Other mild: $n = 5$, severe: $n = 7$.

Between patients with mild disease, South Asians have a significantly elevated antibody titer compared to the Middle Eastern ethnicity groups (Figure 5A). However, the same pattern is not observed between patients with mild disease for epitope coverage, and only the Qatari group has significantly narrower coverage in comparison to South Asians (Figure 5B). Both the Middle Eastern, excluding Qatari, and South Asian groups have significantly higher antibody titers compared to the Qatari group in patients with severe disease (Figure 5A). Interestingly, this trend is not reflected in epitope coverage, where the Middle Eastern group, excluding Qatari, has a significantly broader epitope coverage in comparison to the South Asian group (Figure 5B).

4. Discussion

In the current COVID-19 pandemic, there is increasing interest globally in obtaining a more detailed mechanistic understanding of the underlying immunology of COVID-19 disease at both the B- and T-cell level. A number of papers have described the existence and cross-reactivity of SARS-CoV-2 specific T-cell responses [26–28], as well as correlations with antibody responses [9]. Viral neutralization assays are now providing important new information on neutralizing antibody activity in individuals [29,30], but are typically lower throughput, so reported studies have been on smaller cohorts. Serology assays have thus to date been primarily used in seroprevalence studies to determine the extent of infection in populations, with the rapid serology tests that are typically used in such studies being characterized by qualitative data on single antigens and focusing on simple yes/no answers. Such tests are known to be strongly affected by the time delay between the acute phase of disease and measurement and are not well suited to answer more advanced serological questions such as how the magnitude and breadth of antibody responses varies with time through convalescence, with age or disease severity, or with ethnicity, in large cohorts.

However, with the global roll-out of the first COVID-19 vaccines now well underway, there is increasing interest in how age in particular influences the magnitude and durability of SARS-CoV-2 vaccine responses. In addition, the emergence of SARS-CoV-2 variants of concern, such as the B1.1.7 and B1.351 variants, which appear to allow for at least partial escape from pre-existing antibody responses, necessitates the development of new quantitative, high-throughput serological tools that are suitable to addressing questions about whether, for example, vaccination protects against infection in individuals, or whether (re)-infection can still occur, albeit with reduced disease severity. Quantitative, specific detection of the magnitude and breadth of humoral responses to SARS-CoV-2 antigens seems likely to shed new light on both of these questions.

SARS-CoV-2 encodes a number of major structural proteins that could in principle be used as the basis of next generation serological tests: the nucleocapsid (N), spike (S), envelope (E), and membrane (M) proteins. Recent literature using first generation serology tests suggests that anti-N IgG antibodies are more prevalent than anti-S IgG antibodies in COVID-19 cases and may therefore be better suited to population level studies [31]. However, despite the wealth of available COVID-19 literature, there are few data on anti-E or anti-M antibody responses, implying lesser applicability. Here, we have therefore chosen to focus on gaining a more detailed, quantitative understanding of how antibody responses to the nucleocapsid protein correlate with age, disease severity, and ethnicity.

To enable this, we have engineered a novel, quantitative multi-epitope SARS-CoV-2 protein microarray platform, removing specific nucleocapsid protein epitopes that flanked the structural domains and which were identified as binding strongly and non-specifically to multiple unrelated non-human monoclonal and polyclonal antibodies, yet preserving other more distal, highly discriminatory antibody epitopes in the intrinsically disordered regions. This design resulted in 100% sensitivity and specificity in discrimination of severe COVID-19 cases from pre-pandemic controls in an independent cohort derived from Malaysia. We then utilized this novel immunoassay platform in a cross-sectional multi-ethnic cohort derived from Qatar, consisting of confirmed COVID-19 cases with a gradation of disease severities as well as with a wide age distribution, and have made a number of unexpected observations about age and disease severity influences on the humoral response.

While there is a literature precedent for anti-SARS-CoV-2 antibody titers to increase with disease severity, as also found here in two independent cohorts, we also observed that the breadth of the antibody response—i.e., the number of discrete epitopes recognized per patient—also increased with disease severity (Figure 2), which makes intuitive sense in terms of the amplification of humoral response in individuals with high viral loads and more extensive, longer lasting infection and disease. Notably, the data also suggest that in both independent cohorts, the dominant antigenic epitopes lie in the C-terminal domain of the nucleocapsid protein, with that domain showing more frequent and higher antibody titers (Figure 1) compared to the N-terminal domain in both mild and severe cases. In contrast, antibody recognition of the intrinsically disordered regions appeared to have a lower frequency and lower titer—perhaps suggesting lower affinity of recognition of linear epitopes—supporting the hypothesis that discontinuous epitopes on the surface of the structural domains are the preferred antigenic epitopes on this viral protein and are key to the specificity of this platform.

Classically, older individuals are generally observed to be more susceptible to new infections, due to impairment of adaptive immune responses [32], including immune repertoire exhaustion [33], and deficiency in antigen-driven selection processes [34]. There is also evidence for quite different antibody responses to infection or vaccination in individuals over the age of 50, with differences reported in magnitude and affinity, as well as in antibody class/sub-class, somatic mutation intensity and efficiency, loss of B-cell diversity, and antibody poly-specificity [34–36]. There are thus significant concerns about how well SARS-CoV-2 vaccines will work in older, more vulnerable groups.

Here, disease susceptibility as a function of age in Cohort 3 mirrors expected trends, with adults in the age bracket of 20–40 years being under-represented and those over 50 years being significantly over-represented in the diseased cohort relative to the general population ($p < 0.001$; Table 4). However, unexpectedly, our data show that in Cohorts 2 and 3, both the magnitude and the breadth of anti-SARS-CoV-2 N-protein antibody response increases with age, relative to the under 40 age group, reaching statistical significance in the 51–60 age bracket (Figures 3 and 4), although the small absolute sample numbers in the over 60 age bracket in both cohorts limited the interpretation of our data in that group. This observation might simply reflect increased disease severity in the older age groups, but a trend of increased epitope coverage in the age 51–60 bracket was observed in both mild and severe cases (Figure 4B), arguing that the ability to mount a strong and broad antibody response to SARS-CoV-2 is not compromised by age, at least in these two independent, ethnically diverse cohorts, which is encouraging for the effectiveness of vaccinations in elderly groups. At face value, there appears to be an age cut-off at 60, above which the epitope coverage is lower in Cohort 3, possibly due to impaired adaptive immune responses and/or immune exhaustion in this cohort. However, this is not observed in Cohort 2 and may simply be a function of low sample numbers in that age bracket in Cohort 3. Further research to understand whether the age-related changes observed here in antibody titer and breadth of epitope utilization manifest further in terms of affinity, class/sub-class, effector functions, durability, or poly-specificity of the resultant antibodies will be reported elsewhere.

Table 4. Summary of the demographics of Cohort 3 and the Qatari population. Percentage of each ethnic group in Cohort 3, compared to the percentage of each ethnicity found in the Qatari population. Ethnicities that did not fall under the three broader ethnic groups were excluded from this table (n = 5). Gender distribution in Cohort 3 compared to the gender distribution in the Qatari population. Age distribution in Cohort 3 compared to the age distribution in the Qatari population.

Characteristic	Number of Individuals in Cohort	Percentage of Cohort (%)	Percentage of Qatari Population (%)
	Ethnic Group		
Middle Eastern (Other)	10	10	18.35
Middle Eastern (Qatari)	15	15	10.50
South Asian	70	70	64.32
	Gender		
Male	91	91	72.90
Female	9	9	27.10
	Age Group		
18–40	43	43	69.44
41–50	24	24	19.82
51–60	27	27	7.76
>60	6	6	2.99

The effects of ethnicity on SARS-CoV-2 infection and disease severity remain largely unknown [8]. Data reported by the Centre for Disease Control (CDC) suggest that COVID-19 disproportionally affects certain ethnicities [37]. However, due to other cofounding factors, such as socioeconomic factors and variable access to healthcare, it is challenging to determine whether there is an underlying mechanism to explain the observed disparities in the humoral response between different ethnic groups [8]. Here, amongst the PCR positive group from the Qatar cohort (Cohort 3), we observed significant differences in the magnitude and breadth of antibody responses between the different broad ethnicity groups. The Qatari population as a whole is comprised of ~10% Qataris and ~90% ethnically diverse migrant workers/expats (Table 4 and Supplementary Table S5), the latter of whom can be broadly grouped as being of South Asian, Middle Eastern, or 'Other' ethnicities. The entire Qatar population of ca. 2.8 m people live in a single highly localized geographic region and all have free access to health care, removing one of the confounders referred to above. Our initial expectation therefore was that we might observe a significant difference in antibody responses between individuals as a result of diverse genetic backgrounds or differing susceptibility to severe disease.

All ethnicities in Cohort 3 had higher cumulative reciprocal titers and high epitope coverage scores in severe compared to mild disease, as expected, which reached statistical significance in the non-Qatari Middle Eastern ethnicity group ($p = 0.0045$, reciprocal titers; $p = 0.039$, epitope coverage; Figure 5), but interestingly not in the Qatari group. Unexpectedly, we also observed a significant difference in reciprocal titers between the Middle Eastern (Qatari) and Middle Eastern (non-Qatari) severe disease groups ($p = 0.0078$; Figure 5). It seems reasonable to expect socioeconomic factors to play a role in the incidence of COVID-19 disease in this cohort; notably, females are significantly under-represented in the diseased cohort ($p < 0.01$; Table 4), while there is also significant under-representation of Middle Eastern (non-Qatari) and over-representation of Qatari COVID-19 cases relative to their proportions of the overall population ($p < 0.05$; Table 4), supporting this expectation. However, it is less immediately obvious whether or how socioeconomic factors might affect the humoral response following infection in severe disease cases. Given that the non-Qatari Middle Eastern group comprises nationals from Egypt, Sudan, Syria, Iran, and Yemen (Table S5), it seems possible that genetic differences between the Qatari and non-Qatari Middle Eastern groups might underpin the apparently decreased magnitude of humoral responses following infection and increased risk of COVID-19 disease observed here for the Qatari group. While we did not have access to genome sequence data for this cohort to verify this, it is perhaps relevant that the Qatari population has been reported to have an elevated prevalence of common adult diseases [38], as well as of childhood autoimmune diseases such as type 1 diabetes [39], potentially suggestive of uncharacterized genetic factors that affect humoral immune responses through HLA allelic variation [40].

Amongst the migrant worker groups, we observed a significant difference between the non-Qatari Middle Eastern and South Asian groups, in terms of both reciprocal antibody titers ($p = 0.0013$ for mild disease) and epitope coverage scores ($p = 0.0046$ for severe disease), apparently at least qualitatively further supporting a role for genetic factors and warranting further investigation. Interestingly, the directionality of these comparisons differed between mild and severe disease: reciprocal titers and epitope coverage scores for the non-Qatari Middle Eastern mild disease group were lower than for the South Asian mild disease group, but were higher in the non-Qatari Middle Eastern severe disease group compared to the South Asian severe disease group. This may reflect a greater disease severity in the non-Qatari Middle Eastern group that was not captured by the clinical scores, but more likely again points to intrinsically different humoral responses to SARS-CoV-2 infection amongst the different ethnicity groups in Cohort 3. Further work to explore the underlying basis of these ethnicity-based differences in anti-SARS-CoV-2 humoral responses in a larger cohort, including through HLA allele sequencing, is thus now needed.

Limitations and Further Work

Although this cross-sectional study is statistically powered and identified clear ethnicity- and age-associated differences in both antibody titers and epitope coverage, it is limited by the available cohort sizes, which meant that we were not able to divide the broad ethnic groupings more finely and that certain other ethnicities were essentially absent from the comparisons, while participants over 60 years were under-represented. Furthermore, Cohort 1 comprised convalescent COVID-19 cases with a significantly longer average delay between diagnosis and sample collection, a skewed demographic makeup that is not representative of the general population and with disparate access to healthcare, while Cohort 2 was designed for the case-control validation component of this study, so lacked the spectrum of disease as well as ethnicity data; collectively, these factors limited our ability to integrate results across the three cohorts.

In addition, the study is also limited by its exclusive focus on IgG antibody responses to the nucleocapsid protein. Future studies will expand our quantitative, epitope-resolved antibody assay platform to include the SARS-CoV-2 spike protein and clinically relevant variants thereof; we will also include detection of additional immunoglobulin classes (IgA and IgM) and sub-classes (IgG_{1-4}; IgA_{1-2}), as well as on-array Fc effector function and surrogate neutralization assays, in order to shed further light on the functional consequence of the differential antibody titers observed, particularly in older individuals. Longitudinal studies will enable assessment of the durability of the age-dependent phenomena reported here.

Supplementary Materials: The following are available online at https://www.mdpi.com/article/10.3390/v13050786/s1: Table S1: Oligo pairs used to construct the N-core, NTD, and CTD clones, Table S2: Characteristics of the 17 tiling peptides for the detection of SARS-CoV-2 nucleocapsid phosphoprotein, Table S3: Microarray assay sensitivity as a function of time post onset of symptoms, Table S4: Positive and negative predictor values in Cohort 3, Table S5: Percentage of each ethnicity in Cohort 3, compared to the percentage of each ethnicity found in the Qatari cohort. Ethnicities that did not fall under the three broader ethnic groups were excluded from this table ($n = 5$), Figure S1: Western blot of SARS-2-nucleocapsid structural domains, Figure S2: SARS-CoV-2 N protein amino acid coverage on SARS-CoV-2 microarray, Figure S3: IgG response to SARS-CoV-2 full-length N protein, Figure S4: Linearity of signal as a function of serum dilution on the microarray platform, Figure S5: IgG responses to SARS-CoV-2 N protein variants, Figure S6: Antibody response to peptide 1,Figure S7: Antibody response to peptide 2, Figure S8: Antibody response to peptide 3,Figure S9: Antibody response to peptide 4, Figure S10: Antibody response to peptide 5, Figure S11: Antibody response to peptide 6, Figure S12: Antibody response to peptide 7, Figure S13: Antibody response to peptide 8, Figure S14: Antibody response to peptide 9, Figure S15: Antibody response to peptide 10, Figure S16: Antibody response to peptide 11, Figure S17: Antibody response to peptide 12, Figure S18: Antibody response to peptide 13, Figure S19: Antibody response to peptide 14, Figure S20: Antibody response to peptide 15, Figure S21: Antibody response to peptide 16, Figure S22: Antibody response to peptide 17, Figure S23: IgG response to 6 peptides of the SARS-CoV-2 N protein, Figure S25: Average antibody titer and sensitivity for Cohort 3 patients with mild and severe disease.

Author Contributions: J.M.B., A.A., H.B.A., and O.M.A.E.-A. designed, conceived, and led the study. J.M.B., M.S., A.J.M.N., N.D.A., and T.-M.T. designed the array. A.J.M.N. cloned and expressed the proteins. E.S.M. and W.A.B. provided the samples for Cohort 1. M.S., D.T.S., and S.S.B.-S. developed prototype arrays and assays on Cohort 1. R.M.Z. designed the validation experiment and provided the samples for Cohort 2. N.S.M.R. and K.E. performed the assays on Cohort 2. M.A.Y.A.-N., V.M.-A., and M.A.-M. led the Cohort 3 sample collection, processing, and ethical approvals. H.B.A., N.K.M., N.N.V., A.N., and K.O. optimized the assays on Cohort 3. I.B. and H.B.A. ran the assays on Cohort 3. T.-M.T., M.M., N.H.I., and P.E.M. performed the bioinformatics analysis. J.M.B., A.A., R.M.Z., R.N.R.M., T.-M.T., N.D.A., and H.B.A. interpreted the data. M.M., M.S., H.B.A., and J.M.B. wrote the manuscript. All authors have read and agreed to the published version of the manuscript.

Funding: This research was funded by project funding from QBRI and a grant fund from Hamad Medical Corporation (fund number MRC-05-003). The APC was funded by National Research Foundation, South Africa (grant number 64760).

Institutional Review Board Statement: The study on Cohort 1 was conducted according to the guidelines of the Declaration of Helsinki and approved by the Ethics Committee of the University of Cape Town (UCT; HREC 210/2020; 15 April 2020) and the Ethics Committee of the University of Witwatersrand (M200402; 2 April 2020). Ethical review and approval for the study on Cohort 2 were waived, since it formed part of standard of care at Hospital Sungai Buloh, Selangor, Malaysia. The study on Cohort 3 was conducted according to the guidelines of the Declaration of Helsinki, and approved by the Institutional Review Board Research Ethics Committee of the Hamad Medical Corporation (reference MRC-05-003).

Informed Consent Statement: Written informed consent was obtained from all subjects involved in the study prior to enrolment.

Data Availability Statement: Data are contained within the article or Supplementary Material.

Acknowledgments: We would like to thank all the patients, volunteers, and healthcare co-workers from HMC hospitals. We also acknowledge the help provided by the Anti-Doping Lab-Qatar (ADLQ) and Qatar Red Crescent (QRC) for recruiting control individuals. The research was supported by the Proteomics core facility and Infectious Diseases Interdisciplinary Research Program (ID-IDRP) at QBRI. We acknowledge Nur Izwani Shaiful Bahrin, Noorul Hidayah Badri, and Teh Norleila Abdul Rahman for their assistance in the antibody assay performed at IMR, Malaysia. We would like to thank the Director General of Health Malaysia for his permission to include the validation data in this paper. JMB thanks the National Research Foundation (South Africa) for a SARChI grant.

Conflicts of Interest: A.A., N.D.A., T.-M.T., N.S.M.R., N.H.I., P.M., R.N.R.M., and J.M.B. are employees of Sengenics Corporation, who market a Protein Array product developed in part based on results generated in this study. That product was however not used in this study.

References

1. World Health Organisation. Rolling Updates on Coronavirus Disease (COVID-19). 2020. Available online: https://www.who.int/emergencies/diseases/novel-coronavirus-2019/events-as-they-happen (accessed on 27 April 2021).
2. Khurshid, Z.; Asiri, F.Y.I.; Al Wadaani, H. Human Saliva: Non-Invasive Fluid for Detecting Novel Coronavirus (2019-nCoV). *Int. J. Environ. Res. Public Health* **2020**, *17*, 2225. [CrossRef]
3. Sohrabia, C.; Alsafib, Z.; O'Neilla, N.; Khanb, M.; Kerwanc, A.; Al-Jabirc, A.; Iosifidisa, C.; Aghad, R. World Health Organization declares global emergency: A review of the 2019 novel coronavirus (COVID-19). *Int. J. Surg.* **2020**, *76*, 71–76. [CrossRef]
4. de Walque, D.B.C.M.; Friedman, J.; Gatti, R.V.; Mattoo, A. *How Two Tests Can Help Contain COVID-19 and Revive the Economy*; Research & Policy Briefs; World Bank Group: Washington, DC, USA, 2020; Volume 29. Available online: http://documents.worldbank.org/curated/en/766471586360658318/How-Two-Tests-Can-Help-Contain-COVID-19-and-Revive-the-Economy (accessed on 27 April 2021).
5. World Health Organisation. Coronavirus Disease (COVID-19) Outbreak Situation. 2020. Available online: https://www.who.int/emergencies/diseases/novel-coronavirus-2019 (accessed on 27 April 2021).
6. Escobedo-de la Peña, J.; Rascón Pacheco, R.A.; de Jesús Ascencio-Montiel, I.; González-Figueroa, E.; Fernández-Gárate, J.E.; Medina-Gómez, O.S.; Borja-Bustamante, P.; Santillán-Oropeza, J.A.; Borja-Aburto, V.H. Hypertension, diabetes and obesity, major risk factors for death in patients with COVID-19 in Mexico. *Arch. Med Res.* **2020**. [CrossRef]
7. Ye, C.; Zhang, S.; Zhang, X.; Cai, H.; Gu, J.; Lian, J.; Lu, Y.; Jia, H.; Hu, J.; Jin, C.; et al. Impact of comorbidities on patients with COVID-19: A large retrospective study in Zhejiang, China. *J. Med. Virol.* **2020**, *92*, 2821–2829. [CrossRef]
8. Kopel, J.; Perisetti, A.; Roghani, A.; Aziz, M.; Gajendran, M.; Goyal, H. Racial and Gender-Based Differences in COVID-19. *Front. Public Health* **2020**, *8*. [CrossRef] [PubMed]
9. Reynolds, C.J.; Swadling, L.; Gibbons, J.M.; Pade, C.; Jensen, M.P.; Diniz, M.O.; Schmidt, N.M.; Butler, D.K.; Amin, O.E.; Bailey, S.N.L.; et al. Discordant neutralizing antibody and T cell responses in asymptomatic and mild SARS-CoV-2 infection. *Sci. Immunol.* **2020**, *5*, eabf3698. [CrossRef] [PubMed]
10. Li, D.; Li, J. Immunologic testing for SARS-CoV-2 infection from the antigen perspective. *J. Clin. Microbiol.* **2020**, *59*. [CrossRef]
11. van Tol, S.; Mögling, R.; Li, W.; Godeke, G.-J.; Swart, A.; Bergmans, B.; Brandenburg, A.; Kremer, K.; Murk, J.-L.; van Beek, J.; et al. Accurate serology for SARS-CoV-2 and common human coronaviruses using a multiplex approach. *Emerg. Microbes Infect.* **2020**, *9*, 1965–1973. [CrossRef] [PubMed]
12. de Assis, R.R.; Jain, A.; Nakajima, R.; Jasinskas, A.; Felgner, J.; Obiero, J.M.; Norris, P.J.; Stone, M.; Simmons, G.; Bagri, A.; et al. Analysis of SARS-CoV-2 antibodies in COVID-19 convalescent blood using a coronavirus antigen microarray. *Nat. Commun.* **2021**, *12*, 6. [CrossRef]
13. Zhao, J.; Yuan, Q.; Wang, H.; Liu, W.; Liao, X.; Su, Y.; Wang, X.; Yuan, J.; L, T.; Li, J.; et al. Antibody responses to SARS-CoV-2 in patients of novel coronavirus disease 2019. *Clin. Infect. Dis.* **2020**, *71*, 2027–2034. [CrossRef]

14. Tang, F.; Quan, Y.; Xin, Z.; Wrammert, J.; Ma, M.-J.; Lv, H.; Wang, T.-B.; Yang, H.; Richardus, J.H.; Liu, W.; et al. Lack of Peripheral Memory B Cell Responses in Recovered Patients with Severe Acute Respiratory Syndrome: A Six-Year Follow-Up Study. *J. Immunol.* **2011**, *186*, 7264–7268. [CrossRef]
15. Summary of Probable SARS Cases with Onset of Illness from 1 November 2002 to 31 July 2003. 2003. Available online: https://www.who.int/csr/sars/country/table2004_04_21/en/ (accessed on 27 April 2021).
16. Zhou, W.; Wang, W.; Wang, H.; Lu, R.; Tan, W. First infection by all four non-severe acute respiratory syndrome human coronaviruses takes place during childhood. *BMC Infect. Dis.* **2013**, *13*, 1–8. [CrossRef]
17. Blackburn, J.M.; Shoko, A.; Beeton-Kempen, N. Miniaturized, Microarray-Based Assays for Chemical Proteomic Studies of Protein Function. In *Chemical Genomics and Proteomics: Methods in Molecular Biology*; Humana Press: Totowa, NJ, USA, 2012; Volume 800, pp. 133–162.
18. Adeola, H.A.; Blackburn, J.M.; Rebbeck, T.R.; Zerbini, L.F. Novel potential serological prostate cancer biomarkers using CT100+ cancer antigen microarray platform in a multi-cultural South African cohort. *Oncotarget* **2016**, *7*, 13949–13960. [CrossRef]
19. López-Ratón, M.; Rodríguez-Álvarez, M.X.; Cadarso-Suárez, C.; Gude-Sampedro, F. OptimalCutpoints: An R Package for Selecting Optimal Cutpoints in Diagnostic Tests. *J. Stat. Softw.* **2014**, *61*, 36. [CrossRef]
20. Beeton-Kempen, N.; Duarte, J.; Shoko, A.; Serufuri, J.-M.; John, T.; Cebon, J.; Blackburn, J. Development of a novel, quantitative protein microarray platform for the multiplexed serological analysis of autoantibodies to cancer-testis antigens. *Int. J. Cancer* **2014**, *135*, 1842–1851. [CrossRef]
21. Larkin, J.G. Family History of rheumatoid arthritis—a non-predictor of inflammatory disease? *Rheumatology* **2009**, *49*, 608–609. [CrossRef] [PubMed]
22. Barlow, D.J.; Edwards, M.S.; Thornton, J.M. Continuous and discontinuous protein antigenic determinants. *Nature* **1986**, *322*, 747–748. [CrossRef] [PubMed]
23. Van Regenmortel, M.H.V. Mapping Epitope Structure and Activity: From One-Dimensional Prediction to Four-Dimensional Description of Antigenic Specificity. *Methods* **1996**, *9*, 465–472. [CrossRef]
24. Chapman-Smith, A.; Cronan, J.E., Jr. Molecular Biology of Biotin Attachment to Proteins. *J. Nutr.* **1999**, *129*, 477S–484S. [CrossRef] [PubMed]
25. Ahmed, S.F.; Quadeer, A.A.; McKay, M.R. Preliminary Identification of Potential Vaccine Targets for the COVID-19 Coronavirus (SARS-CoV-2) Based on SARS-CoV Immunological Studies. *Viruses* **2020**, *12*, 254. [CrossRef]
26. Grifoni, A.; Weiskopf, D.; Ramirez, S.I.; Mateus, J.; Dan, J.M.; Moderbacher, C.R.; Rawlings, S.A.; Sutherland, A.; Premkumar, L.; Jadi, R.S.; et al. Targets of T Cell Responses to SARS-CoV-2 Coronavirus in Humans with COVID-19 Disease and Unexposed Individuals. *Cell* **2020**, *181*, 1489–1501.e1415. [CrossRef]
27. Peng, Y.; Mentzer, A.J.; Liu, G.; Yao, X.; Yin, Z.; Dong, D.; Dejnirattisai, W.; Rostron, T.; Supasa, P.; Liu, C.; et al. Broad and strong memory CD4+ and CD8+ T cells induced by SARS-CoV-2 in UK convalescent individuals following COVID-19. *Nat. Immunol.* **2020**, *21*, 1336–1345. [CrossRef]
28. Sekine, T.; Perez-Potti, A.; Rivera-Ballesteros, O.; Strålin, K.; Gorin, J.B.; Olsson, A.; Llewellyn-Lacey, S.; Kamal, H.; Bogdanovic, G.; Muschiol, S.; et al. Robust T Cell Immunity in Convalescent Individuals with Asymptomatic or Mild COVID-19. *Cell* **2020**, *183*, 158–168.e114. [CrossRef]
29. Cele, S.; Gazy, I.; Jackson, L.; Hwa, S.-H.; Tegally, H.; Lustig, G.; Giandhari, J.; Pillay, S.; Wilkinson, E.; Naidoo, Y.; et al. Escape of SARS-CoV-2 501Y.V2 from neutralization by convalescent plasma. *Nature* **2021**. [CrossRef]
30. Wibmer, C.K.; Ayres, F.; Hermanus, T.; Madzivhandila, M.; Kgagudi, P.; Oosthuysen, B.; Lambson, B.E.; de Oliveira, T.; Vermeulen, M.; van der Berg, K.; et al. SARS-CoV-2 501Y.V2 escapes neutralization by South African COVID-19 donor plasma. *Nat. Med.* **2021**, *27*, 622–625. [CrossRef]
31. Manisty, C.; Treibel, T.A.; Jensen, M.; Semper, A.; Joy, G.; Gupta, R.K.; Cutino-Moguel, T.; Andiapen, M.; Jones, J.; Taylor, S.; et al. Time series analysis and mechanistic modelling of heterogeneity and sero-reversion in antibody responses to mild SARS-CoV-2 infection. *EBioMedicine* **2021**, *65*, 103259. [CrossRef] [PubMed]
32. Castelo-Branco, C.; Soveral, I. The immune system and aging: A review. *Gynecol. Endocrinol.* **2014**, *30*, 16–22. [CrossRef]
33. Gibson, K.L.; Wu, Y.-C.; Barnett, Y.; Duggan, O.; Vaughan, R.; Kondeatis, E.; Nilsson, B.-O.; Wikby, A.; Kipling, D.; Dunn-Walters, D.K. B-cell diversity decreases in old age and is correlated with poor health status. *Aging Cell* **2009**, *8*, 18–25. [CrossRef] [PubMed]
34. Martin, V.; Wu, Y.-C.; Kipling, D.; Dunn-Walters, D. Ageing of the B-cell repertoire. *Philos. Trans. R. Soc. B Biol. Sci.* **2015**, *370*, 20140237. [CrossRef] [PubMed]
35. Goodwin, K.; Viboud, C.; Simonsen, L. Antibody response to influenza vaccination in the elderly: A quantitative review. *Vaccine* **2006**, *24*, 1159–1169. [CrossRef]
36. Davydov, A.N.; Obraztsova, A.S.; Lebedin, M.Y.; Turchaninova, M.A.; Staroverov, D.B.; Merzlyak, E.M.; Sharonov, G.V.; Kladova, O.; Shugay, M.; Britanova, O.V.; et al. Comparative Analysis of B-Cell Receptor Repertoires Induced by Live Yellow Fever Vaccine in Young and Middle-Age Donors. *Front. Immunol.* **2018**, *9*. [CrossRef] [PubMed]
37. Smith, A.R.; DeVies, J.; Caruso, E.; Radhakrishnan, L.; Sheppard, M.; Stein, Z.; Calanan, R.M.; Hartnett, K.P.; Kite-Powell, A.; Rodgers, L.; et al. Emergency Department Visits for COVID-19 by Race and Ethnicity—13 States, October–December 2020. *MMWR Morb. Mortal. Wkly. Rep.* **2021**, *70*, 566–569. [CrossRef] [PubMed]
38. Bener, A.; Hussain, R.; Teebi, A.S. Consanguineous marriages and their effects on common adult diseases: Studies from an endogamous population. *Med. Princ. Pract.* **2007**, *16*, 262–267. [CrossRef] [PubMed]

39. De Sanctis, V. Type 1 and Type 2 diabetes in children and adolescents: A public health problem in Qatar. The experience of Pediatric Diabetes Center at Hamad General Hospital (HGH) of Doha. *Acta Biomed.* **2018**, *89*, 5–6. [CrossRef]
40. Al Naqbi, H.; Mawart, A.; Alshamsi, J.; Al Safar, H.; Tay, G.K. Major histocompatibility complex (MHC) associations with diseases in ethnic groups of the Arabian Peninsula. *Immunogenetics* **2021**, *73*, 131–152. [CrossRef] [PubMed]

MDPI

Review

A Review of COVID-19 Modelling Strategies in Three Countries to Develop a Research Framework for Regional Areas

Azizur Rahman [1,2,*], Md Abdul Kuddus [1,3,4], Ryan H. L. Ip [1] and Michael Bewong [1]

1 School of Computing, Mathematics and Engineering, Charles Sturt University, Wagga Wagga, NSW 2678, Australia; mdabdul.kuddus@my.jcu.edu.au (M.A.K.); hoip@csu.edu.au (R.H.L.I.); mbewong@csu.edu.au (M.B.)
2 Institute for Land, Water and Society (ILWS), Charles Sturt University, Albury, NSW 2640, Australia
3 Australian Institute of Tropical Health and Medicine, James Cook University, Townsville, QLD 4814, Australia
4 Department of Mathematics, University of Rajshahi, Rajshahi 6205, Bangladesh
* Correspondence: azrahman@csu.edu.au

Abstract: At the end of December 2019, an outbreak of COVID-19 occurred in Wuhan city, China. Modelling plays a crucial role in developing a strategy to prevent a disease outbreak from spreading around the globe. Models have contributed to the perspicacity of epidemiological variations between and within nations and the planning of desired control strategies. In this paper, a literature review was conducted to summarise knowledge about COVID-19 disease modelling in three countries—China, the UK and Australia—to develop a robust research framework for the regional areas that are urban and rural health districts of New South Wales, Australia. In different aspects of modelling, summarising disease and intervention strategies can help policymakers control the outbreak of COVID-19 and may motivate modelling disease-related research at a finer level of regional geospatial scales in the future.

Keywords: COVID-19; models; different settings; intervention strategies; NSW

Citation: Rahman, A.; Kuddus, M.A.; Ip, R.H.L.; Bewong, M. A Review of COVID-19 Modelling Strategies in Three Countries to Develop a Research Framework for Regional Areas. *Viruses* **2021**, *13*, 2185. https://doi.org/10.3390/v13112185

Academic Editors: Burtram C. Fielding and Georgia Schäfer

Received: 25 August 2021
Accepted: 26 October 2021
Published: 29 October 2021

1. Introduction

Over the last few decades, the world faced a massive challenge in controlling infectious disease outbreaks in several areas [1]. Recently, a new infectious disease, SARS-CoV-2 named COVID-19, a virus of *coronaviridae* family and genus beta coronavirus, has emerged globally, and almost all countries and territories are now fighting against this newly appeared infectious disease [2]. The Municipal Commission in Wuhan, China, reported a cluster of pneumonia cases that had an unfamiliar etiology on 12th December 2019. COVID-19 was first identified in Wuhan city, Hubei Province of China, on 31st December 2019, and it spread so fast that within only five months, nearly two million people were infected in 185 countries around the world [3]. On 11th March 2020, the World Health Organization (WHO) announced the transmission of COVID-19 as a global pandemic because of the rapid increment of its infection rate [4]. Following SARS-CoV, which originated in China in 2003, and MERS-CoV, which originated in Saudi Arabia in 2013, SARS-CoV-2 seems to have become the third most significant public health concern of its type. The current fatality rate for COVID-19 cases is about 3.4%, significantly less than SARS and MERS, but potentially higher than those reported for endemic human non-SARS CoV infections [5].

The number of cases quickly rose to 44, with 11 of these patients in severe condition on 3rd January 2020. The COVID-19 virus spread across mainland China with over 30 thousand confirmed cases and over 600 deaths within only one month [6]. The World Health Organization (WHO) published an online resource that presented countries with guidance on detecting, testing and controlling possible cases on 10th January 2020 [7]. The first case outside of China was reported on 13 January 2020. Then, by 11th March 2020, the WHO declared COVID-19 to be a pandemic, based on its fast spread outside China.

As of 11th November 2020, over 51.3 million people have been infected globally, with a 2.5% death rate [8]. Currently, almost 47.7% of the total global infections are in three countries—the United States (US), India, and Brazil. Together, deaths in these countries make up around 41.7% of global deaths [8]. According to the Worldometer estimation, up to the date 20th July 2021, nearly 191.7 million people have been identified as infected, with more than 4 million deaths, and about 174.5 million individuals have recovered in 213 countries and territories around the globe [9].

In the US, state and local governments, following the Center for Disease Control (CDC) guidance, started monitoring all individuals who had been in close proximity with confirmed COVID-19 cases. As a result, by 26th February 2020, 12 travel-related positive cases and three positive cases with no travel history were documented [10]. Specifically, the latter category of infections was a cause for concern since it indicated a significantly higher presence of the virus in the United States. In worldwide COVID-19 deaths, the US has been severely burdened by the disease and it alone accounts for about 18.9% of the global deaths, followed by Brazil and India with about 12.8% and 10.0% of global deaths, respectively [4].

The first cases of COVID-19 were linked to a live animal market in Wuhan, China [11]; however, the current rapid spread is via human-to-human transmission. Once infected, the individual will first undergo a period without visible clinical symptoms, called a latent SARS-CoV-2 infection. People with latent SARS-CoV-2 can become infectious one to two days before the onset of symptoms and continue to be infectious up to seven days after that [12]. Therefore, after a certain period, the latent SARS-CoV-2 infection progresses to an active COVID-19 infection. The disease spreads quickly from a person with active COVID-19 infection to another person when the infectious and susceptible persons are close [13]. The spread of COVID-19 depends on the length of exposure of susceptible people to the infected person [14]. It is, in turn, dependent on many factors, such as the crowdedness of the environment, any super-spreading events, the prevailing climatic conditions and the immune status of the exposed individual [15].

Despite extensive epidemiological research on various coronaviruses, there are still many unknowns about this new disease. It is thought that COVID-19 primarily spreads via respiratory droplets and aerosol and has an incubation period of up to 14 days, with symptom onset generally occurring at around days 5–6, similar to SARS-CoV, the cause of the severe acute respiratory syndrome (SARS) epidemic in 2002 [16–18]. However, unlike SARS-CoV, which resulted in high viral loads in the lower respiratory tract and led to viral shedding with symptom onset, SARS-CoV-2 has been shown to result in viral shedding due to asymptomatic infection from the upper respiratory tract and making it problematic to organisation preventative procedures that depend on symptomatology [6,19]. As a result, it led to an extreme contact rate from infectious persons to susceptible individuals, and while SARS was basically under control within eight months, the nature of COVID-19 is resembled differently due to the several variants [20]. COVID-19 has various signs and symptoms, varying from a mild cough and fever to a shortness of breath, pain, and even anosmia [21]. The disease is also severely prevalent, with most affected persons being asymptomatic or presenting only mild symptoms. However, the other critical forms of COVID-19 require hospitalisations and, in many cases, prolonged intubations. Treatment for the COVID-19 generally focused on supportive capacities, with only limited antiviral medicine (and announced vaccines in all nations that are open or ready for extensive use to remarkably reduce the number of people dying from COVID-19 through vaccination), presenting some promise at that moment [22–26].

A recent study on risk factors conducted by the Oxford Royal College of General Practitioners Research and the Surveillance Centre primary care network investigated severe disease combined infection rate and disease rate and showed a higher probability of infection for older people, men, people of ethnicity other than white, as well as people from areas with a higher socio-economically deprivation or population density [27]. In addition, initial studies showed COVID-19 to be associated with older age, ethnicity, high population density, and comorbidities such as respiratory infections, hypertension, diabetes, and cardiovascular diseases [19,21,28–30]. Notwithstanding significant improvements in science and technology, our perception of the pathogenesis of COVID-19 still seems to be rudimentary, with new (and sometimes conflicting) data emerging almost daily to address the pandemic more efficiently and a race to possible intervention strategy selections.

Modelling has been used as a tool to address gaps in knowledge and to inform health policies for the prevention and control of COVID-19 [31–34]. Currently, researchers have developed different types of modelling approaches to estimate the relationship between COVID-19 and various risk factors in different sociodemographic and geospatial settings [21,33,35–37]. In addition, modelling studies also explore the impact of different intervention strategies to identify the most effective ones. In this study, we carry out a literature review on COVID-19 and infectious disease modelling strategies to develop a robust research framework for the regional areas of New South Wales (NSW), Australia. We believe this may help improve the control strategy for COVID-19 epidemics at the regional level in NSW, and the prospective modelling outcomes will be helpful to decision-makers.

2. Modelling Experience from Three Countries for COVID-19

In this section, we appraise different modelling strategies used for the COVID-19 outbreak in three countries—China, the UK, and Australia. Within Australia, we will focus on the transmission dynamics modelling approach considered in NSW.

2.1. Models with Single and Multiple Interventions

A mathematical model is an essential tool to determine which combination interventions would be most effective for reducing the outbreak of COVID-19. Prem et al. [34] developed a modified SEIR model to investigate the impact of physical distancing and population mixing on the progression of COVID-19 in Wuhan, China. In this study, the authors applied synthetic location-specific contact patterns in Wuhan and adjusted these for school closures, extended workplace closures, and decreasing mixing in the general population. They also considered predicting the impact of lifting control measures by permitting people to return to work in their offices. This study found that physical distancing measures were the most useful for controlling COVID-19 in Wuhan. However, implementing physical distancing measures produced varying results, with the duration of infectiousness and the adaptation of school and workplace closures during COVID-19 outbreaks. This study suggests that the premature and sudden lifting of restrictions could lead to a secondary outbreak. Nevertheless, the risk of a secondary outbreak could be minimised or controlled by relaxing restrictions systematically. The limitations of this study are statistical uncertainties about measures of the basic reproduction number and the continuation of infectiousness.

Most of the mathematical modelling studies focus on the transmission dynamics of COVID-19 and do not consider the changing epidemiology and temporal and spatial transmission heterogeneity. Hou et al. [38] developed a modified multi-stage SEIR model to describe the transmission dynamics of COVID-19 in Wuhan at different spatio-temporal scales. In this study, the authors consider the variation in infectivity and introduce the control, the basic reproduction number, by assuming the exposed population to be infectious and simulate the future spread of COVID-19 across Wuhan. The authors also built a novel source-tracing algorithm to infer the initial exposed number of individuals and to estimate the number of infections during the epidemic. The significant findings of this study are that the spatial patterns of COVID-19 spread are heterogeneous, and the infectivity is significantly more remarkable for the exposed population than the infectious population. However, in this study, the predicted exposed population is much greater than the officially reported size of the infectious population in Wuhan.

Due to the insufficient number of COVID-19 vaccines in the early stage, in many countries, lockdown is one of the most effective measures to control the spread of infection and to evaluate the influence of non-pharmaceutical interventions, including the reopening of schools and workplaces, as well as household contacts, and the broader relaxation of physical distancing. Panovaka-Griffiths et al. [39] develop a stochastic individual-based model for the transmission dynamics of COVID-19 in the UK to estimate the impact of school reopening strategies and contact tracing–testing scenarios. The results showed that increasing testing levels and effective contact tracing coupled with isolation might control COVID-19 in the UK. However, without raising testing levels and widespread contact tracing, the reopening of schools together with the gradual relaxing of lockdown measures are likely to cause secondary outbreaks of COVID-19. This study suggests that for preventing secondary spikes in COVID-19 in the UK, the relaxation of physical distancing such as the reopening of schools must be followed by large-scale, effective contact tracing, supported by isolation and the testing of symptomatic individuals [39].

Despite the first confirmed case of COVID-19 in the UK occurring on 30th January 2020, the UK government waited until lab-confirmed cases reached 11,080 before initiating a lockdown on 24th March [40]. How and when to make public health decisions during epidemics are challenging questions to answer. The appropriate policy response should be based on scientific evidence, which depends on good data and modelling. Modelling is the most effective way of measuring and controlling the current outbreak of COVID-19. The critical parameter for explaining the spread of COVID-19 is the basic reproduction number, which is the expected number of secondary cases caused by a single infectious individual introduced into a susceptible population. If the basic reproduction number is less than one, the disease is endemic; otherwise, it is an epidemic. In looking at the effect of the basic reproduction number on the dynamics of the outbreak of COVID-19 in the UK, Wang et al. [41] considered the SIR and SEIR model. Here, the authors defined four types of populations; susceptible (S)—those who are not in contact with the virus but might be infected as a result of transmission from an infected individual; Exposed (E)—those who are infected but not infectious; Infected (I)—those who are infected and infectious; Removed (R)—those who were previously infected but are now free of the disease. The results showed that the basic reproduction number plays a crucial role in explaining the dynamics of the outbreak of COVID-19 in the UK, but due to the novel nature of COVID-19, there is still a challenge to evaluate the epidemiological implications. Therefore, further research is urgently required to fill the gaps.

COVID-19 spreads quickly from one person with the virus to another person when the infectious person coughs and the susceptible person comes into physical contact [13]. Stutt et al. [42] developed a mathematical model to show the effect of wearing facemasks with or without lockdown times on the transmission dynamics of COVID-19 in the UK. The results showed that when the public adopts wearing facemasks most of the time, the effective reproduction number can be reduced to below one, leading towards the elimination of epidemic spread. Furthermore, when lockdown times are implemented in combination with facemask use, there is a lesser spread of the disease, and the secondary peak is not as high. This study suggested that a combination of strategies, including wearing facemasks and social distancing or lockdowns, may constitute a satisfactory policy for controlling COVID-19.

COVID-19 has placed significant extra pressure on hospital intensive care services in many countries, including Australia [43]. Mathematical modelling can provide important insights into the likely cause of the epidemic—these insights are valuable for the intensive care services during the epidemic. Adekunle et al. [36] developed a stochastic metapopulation model to describe the effect of travel bans imposed globally and within Australia on international flight travel volumes. The results showed that travel bans on international passengers arriving from different countries, including Iran, Italy and South Korea, had no significant impact on decreasing the outbreak of COVID-19 cases. However, in the case of a ban on travellers from China, it did have a significant impact. The authors mentioned that one reason for this was that the prevalence of the disease in countries like Iran, Italy and South Korea was lower than in China, and Italy had previously implemented a lockdown by the time Australia implemented restrictions on travellers coming from Italy. Thus, they suggested that the travel ban is very efficient in delaying the extensive transmission of COVID-19. A similar conclusion was drawn by Ip et al. [44] who evaluated various mitigation policies implemented by the state and federal governments of Australia using a generalised space–time autoregressive model. They found that both international and interstate border controls helped to reduce the number of new COVID-19 cases in Australia.

Kang et al. [6] explained the spatio-temporal pattern and explored the spatial relationship of the COVID-19 epidemic in mainland China. This study found that most of the models, except medical-care-based connection models, showed a significant spatial relationship of COVID-19 infections, which means that the management of the spatial spread in the early stage of COVID-19 is very significant for the control of the further transmission. However, although this study has incorporated the spatial aspect of COVID-19, it has some limitations. Firstly, this study did not take into account the number of suspected cases. Therefore, it is a challenge to understand the spatio-temporal transmission of COVID-19. Secondly, this study did not incorporate the urban–rural connection, which might have an important impact on transmission. Therefore, further research is needed to include the most critical factors and to explore the spatial spread of COVID-19.

Costantino et al. [45] developed a deterministic model to further explore the effectiveness of full and partial travel bans in Australia for travellers from China against the spread of COVID-19. They modelled three basic scenarios—no ban, the current ban, followed by a full or partial lifting to examine the influence of travel bans on the dynamics of COVID-19 outbreak control. Moreover, they used COVID-19 incidence data from China and details of passenger flights between China and Australia during and after the outbreak in China, obtained from incoming passenger arrival record cards. The results show that without a travel ban, an increase of more than 2000 cases and around 400 deaths would have occurred. The complete travel ban decreased the number of cases by more than 86%, while the partial travel ban reduced the number of cases by 50%. These figures indicate the efficacy of policy decisions. This study suggests that imposing travel restrictions with a country (China) experiencing an epidemic peak is highly effective. Further tabulated information of the key literature review on COVID-19 modelling in China, the UK, and Australia is summarised in Table 1, which follows.

Table 1. Summary of the key findings of some important literature about COVID-19 modelling in China, the UK, and Australia.

Countries	Author(s)	Research Aims	Methodology	Significant Findings	Strategies
China	Zhao and Chen [46]	To characterise the dynamics of COVID-19 and explicitly parameterise the intervention effects of control measures in China.	A Susceptible Un-quarantined, Quarantined infected, Confirmed infected (SUQC) model is applied to analyse the daily cases of COVID-19 outbreak in China.	The quarantine and control measures are effective in preventing the spread of COVID-19.	Quarantine and control measures.
China	Liu et al. [47]	To summarise and share the experience of controlling the spread of COVID-19 and provide effective recommendations to enable other countries to save lives.	A modified SEIR model is applied. It considered many influencing factors including spring festival, sealing off the city and construction of the fangcang shelter hospital.	Four different scenarios were investigated to capture different intervention practices. The combination of intervention measures is the only effective way to control the spread and not a single one of them can be omitted.	Seal off the city, enough medical resources, a combination of several interventions, authorities did nothing to control the epidemic.
China	Hao et al. [48]	To reconstruct the full-spectrum dynamics of COVID-19 between 1 January 2020 and 8 March 2020 across five periods marked by events and interventions based on 32,583 laboratory confirmed cases.	A modified susceptible-exposed-presymtomatic infectious-ascertained infectious-unascertained infectious-isolated-removed (SAPHIRE) SEIR model is applied and considered presymtomatic infectiousness, time-varying ascertainment rate, transmission rates and population movements.	Identified two key features of the outbreak: high covertness and high transmissibility. Found multi-pronged interventions had considerable positive effects on controlling the outbreak of COVID-19 and decreasing the reproduction number.	Presymtomatic infectiousness, time-varying ascertainment rate, transmission rates and population movements.
China	Wu et al. [49]	To estimate the clinical age specific severity, which requires properly adjusting for the case ascertainment rate and the delay between the onset of symptoms and death.	A SIR model is applied, which included the number of passengers and confirmed cases who returned to their countries from Wuhan on chartered flights.	Estimated the overall case, symptomatic case, fatality risk, and found that the risk of symptomatic infection increased with age.	Case ascertainment rate, symptoms onset and deaths.

Table 1. *Cont.*

Countries	Author(s)	Research Aims	Methodology	Significant Findings	Strategies
China	Mizumoto et al. [50]	To investigate a link between the wet market and the early spread of COVID-19 in Wuhan, China.	A quantitative modelling framework was applied, which includes daily series of COVID-19 incidence to estimate the reproduction number for market to human and human to human transmission, the probability of reporting and the early effects on public health.	Found that the basic reproduction number of market to human transmission was lower than for human to human transmission. In contrast, the reporting rate for cases stemming from market to human transmission is 2–34 fold higher than that for cases stemming from human to human transmission, suggesting that contact history with the wet market plays an important role in identifying COVID-19 cases.	Wet market to human and human to human transmission.
China	Zhang et al. [51]	To analyse contact survey data for Wuhan and Shanghai before and during the outbreak and contact-tracing information from Hunan province.	A simple SIR model applied to show the impact of age, contact patterns, social distancing, susceptibility to infection for the dynamics of COVID-19 in Hunan province, China.	The results showed that children 0 to 14 years of age are less susceptible to COVID-19 infection than adults 15 to 64 years of age. However, individuals 65+ years of age are more susceptible to infection. Further, this study found that social distancing alone is sufficient to control COVID-19 in China.	Age, contact patterns, social distancing and susceptibility to infection.
China	Pang et al. [52]	To compute the basic reproduction number and analyse the disease free equilibrium as well as sensitivity analysis.	A modified SEIR model was used to explore the dynamics of COVID-19 in Wuhan, China and calculate the most important parameters.	The transmission rate is the most important parameter that can increase the severity of COVID-19 outbreak.	Transmission rate.
UK	Yang et al. [53]	To conduct a feasibility study for robustly estimating the number and distribution of infection, growth of death, peaks and lengths of COVID-19 breakouts by taking multiple interventions in the UK.	A modified SEIR model is used to infer the impact of mitigation, suppression and multiple rolling interventions for controlling the COVID-19 outbreak in the UK.	Rolling intervention is probably an optimal strategy to effectively and efficiently control COVID-19 outbreaks in the UK.	Mitigation, suppression.

Table 1. *Cont.*

Countries	Author(s)	Research Aims	Methodology	Significant Findings	Strategies
UK	Davies et al. [54]	To assess the potential impact of different control measures for mitigating the burden of COVID-19 cases in the UK.	A stochastic age-structured transmission dynamic model is applied to explore the range of intervention scenarios and estimate the impact of varying adherence to interventions across countries.	Four base interventions including school closures, physical distancing, shielding of people aged 70 years or older and self-isolation were each likely to decrease the basic reproduction number but not sufficiently to prevent ICU demand from exceeding health service capacity. Intensive interventions with lockdown periods will need to be considered to prevent excessive health-care demand.	School closures, physical distancing, shielding of people aged 70 years or older and self-isolation.
UK	Booton et al. [55]	To develop a regional transmission dynamics model of COVID-19, for use in estimating the number of infections, deaths and required acute and intensive care (IC) beds in the south west of the UK.	A modified age-structured SEIR model to estimate cumulative cases and deaths and the impact of interventions.	Before any interventions, the basic reproduction number value is 2.6, with social distancing reducing this value to 2.3 and lockdowns/school closures further reducing the basic reproduction number to 0.6, which indicates that lockdowns/school closures are very effective interventions for controlling COVID-19.	Social distancing, lockdowns/school closures.
UK	Stutt et al. [43]	To estimate the impact of facemasks as a non-pharmaceutical intervention, especially in the setting where high-technology interventions including contact tracing or rapid case detection are not feasible.	A modified SEIR model is used to examine the dynamics of COVID-19 epidemics when facemasks are worn by the public, with or without imposed lockdowns.	The results revealed that when facemasks are used by the public all the time, the effective reproduction number can be decreased below 1, leading to the mitigation of epidemic spread. Further, with the combination of lockdowns and 100% facemask use, there is vastly less disease spread.	Lockdowns and facemasks.
UK	Rawson et al. [56]	To investigate the efficacy of two potential lockdown release strategies including ending quarantine and a re-integration approach.	A SEIR model is used to explore the gradual release strategy by allowing different fractions of lockdown.	Ending quarantine for the entire population simultaneously is a high-risk strategy; a gradual re-integration approach would be more reliable.	Lockdowns.

Table 1. *Cont.*

Countries	Author(s)	Research Aims	Methodology	Significant Findings	Strategies
UK	Thompson [57]	To predict the effects of different non-pharmaceutical interventions.	A simple SIR model is used to demonstrate the principle that a reduction in transmission can delay and reduce the height of the epidemic peak under different non-pharmaceutical interventions.	The results revealed that lockdowns are more effective than other non-pharmaceutical interventions and need to be implemented immediately for controlling COVID-19 in the UK.	Lockdowns, school closures, social distancing, shielding of high-risk individuals and self-isolation.
UK	Peiliang and Li [58]	To predict the number of cases and estimate the basic reproduction number under different scenarios.	A modified SEIR model structure is used to explore the effect of time lag and the probability distribution of model states under different interventions.	Self-isolation can reduce the basic reproduction from 7 to 2 in the UK. Strict lockdowns and social distancing are effective interventions for reducing the basic reproduction number below 2.	Self-isolation, lockdowns and social distancing.
Australia	Chang et al. [59]	To compare several intervention strategies including restrictions on international travel, case isolation, home quarantine, social distancing and school closures.	An agent-based model is developed for a fine-grained computational simulation of the ongoing COVID-19 pandemic in Australia.	The results showed that school closures do not bring decisive benefits unless coupled with a high level of social distancing. Furthermore, a 90% level of social distancing is effective to control the COVID-19 within 13–14 weeks when coupled with effective case isolation and international travel restrictions.	International travel, case isolation, home quarantine, social distancing and school closures.
Australia	Fox et al. [60]	To explore the effect of varying the infection reproduction number, which can be reduced by effective social distancing measures at the peak of the epidemic.	A simple SEIR model is used, which includes household quarantine and social distancing.	The results showed that without social distancing, the number of people requiring hospitalisation in NSW will peak at 450 per 100,000 population and the number of individuals requiring critical care are at 150 per 100,000 population.	Household quarantine and social distancing.
Australia	Moss et al. [61]	To estimate the healthcare requirements for COVID-19 patients in the context of broader public health measures.	An age- and risk-stratified transmission model of COVID-19 infection is used to simulate an unmitigated epidemic in current estimates of transmissibility and severity.	The results showed that case isolation and contact quarantine alone will not be sufficient to constrain case presentations within a feasible level of expansion of health sector capacity. Social restrictions will need to be applied at some level during the epidemic.	Case isolation and contact quarantine.

Table 1. *Cont.*

Countries	Author(s)	Research Aims	Methodology	Significant Findings	Strategies
Australia	Milne and Xie [62]	To evaluate a range of social distancing measures and to determine the most effective strategies to reduce the peak daily infection rate and consequential pressure on the healthcare system.	A transmission dynamics individual-based model is used to generate the rate of growth in cases, the magnitude of the epidemic peak and the outbreak duration.	The application of all four social distancing interventions including school closures, workplace non-attendance, increased case isolation and community contact reduction is highly effective for controlling COVID-19 in Australia.	School closures, workplace non-attendance, increased case isolation and community contact.
Australia	Costantino et al. [45]	To test the impact of travel bans on epidemic control in Australia.	An age-specific deterministic model is used to explore the impact of three travel ban scenarios.	The results showed that without travel bans the epidemic in Australia will continue for more than a year, partial travel is minimal and may be a policy option. Finally, travel restrictions are highly effective for controlling the outbreak of COVID-19 in Australia.	Travel restrictions.
Australia	Adekunle et al. [36]	To evaluate the effect of travel bans in the Australian context and predict the epidemic until May 2020.	A stochastic meta-population model was used. It categorises the global population into susceptible, exposed, infectious or recovered (SEIR) individuals.	The results showed that without travel bans Australia would have experienced local transmission as early as January 15 and possibly would have become the Pacific epicentre. Furthermore, having interventions in place can reduce the outbreak of local transmissions of COVID-19 in Australia.	Travel bans.
Australia	Price et al. [63]	To describe how the epidemic and public health response unfolded in Australia up to 13 April 2020.	A SEEIIR model is applied to estimate the time-varying effective reproduction number, which can be used for controlling COVID-19 in Australia.	The results showed that the effective reproduction number is likely below 1 in each Australian state since mid-March and forecast that hospital ward and intensive care unit occupancy would remain below capacity thresholds during the last two weeks of March.	Intensity and timing public health intervention.

2.2. Models with Age Structure and Vaccination

Age is one of the significant factors which can influence the occurrence and severity of the COVID-19 disease. Chang et al. [59] developed an agent-based model for transmission dynamics of the ongoing COVID-19 outbreak in Australia. The authors applied the model to compare several intervention strategies, including travel restrictions, case isolation, school closures, social distancing, and home quarantine. The results showed that the rate of symptomatic cases in children is one-fifth of the rate for adults. This study also shows that the intervention of school closures alone was not effective unless coupled with a high level of social distancing. Therefore, the authors asserted that the combination of social distancing with effective isolation and international travel restrictions was the most effective way to control the outbreak of COVID-19.

Vaccination is often considered the best way to prevent or control outbreaks of infectious diseases including COVID-19 [64]. In addition, in the cases of some infectious diseases, there is no specific treatment except vaccination. Although the exploration of vaccines for COVID-19 was a great challenge, different types of vaccines are now available to combat the spread of COVID-19. The European Medicines Agency and the Italian Medicines Agency have approved Pfizer, Moderna, AstraZeneca AZD1222 and J&J Ad26.COV2.S on 13th March 2021 [65].

Table 2 presents a tabulated summary of the current models that include the vaccination strategies specifically focused on China, the UK, and Australia. For instance, McBryde et al. (2021) developed a COVID-19 model with a vaccination to explore the direct and indirect effects of vaccination by vaccine type, age strategy, and coverage in Australia [66]. The model incorporated some crucial factors, including age-specific mixing, infectiousness, susceptibility and severity, to examine the epidemic under different intervention scenarios. The authors found that the current mixed program, including vaccination with AstraZeneca and Pfizer, would not achieve herd immunity unless 85% of Australia is covered, including 5–16 years of age and considering the effective reproduction number for Delta variant is 5. However, when the value of the effective reproduction number is 3, the mixed program can achieve herd immunity at 60–70% coverage without vaccinating 5–15 years of age. The general finding of this study was that vaccination can prevent 85% of death compared to without vaccination [66].

In 2021, with numerous vaccines becoming available in Australia, Maclntyre et al. (2021) developed a compartmental COVID-19 model to explore the vaccine's effectiveness for target groups, including health workers, young people and older adults, and mass vaccination in NSW Australia [67]. For the target group, results showed that health worker vaccination is necessary for health system resilience. Furthermore, age-based policies with restricted doses of the vaccine can reduce a small amount of infections, but vaccinating older people reduces the prevalence of death. On the other hand, mass vaccination, including 66% of the NSW population, can achieve herd immunity. However, this study also found that slower vaccination rates can lead to a prolonging of the COVID-19 pandemic, and a higher number of cases and deaths in the population [67].

Table 2. Some current models that include vaccination strategies in China, the UK, and Australia.

Countries	Author(s)	Model	Assumptions Implicit (and Explicit)	Applications in Predicting COVID-19	Policy Implications
Australia	McBryde et al. [66]	An individual based model with vaccination.	The model incorporates some important factors including age-specific mixing, infectiousness, susceptibility, and severity to examine the epidemic size under different intervention scenarios.	Predicting the impact of combination second doses vaccination strategies including AstraZeneca and Pfizer. Vaccination can prevent 85% of death compared with no vaccination.	Australia government can take immediate action to vaccinate all population.
Australia	MacIntyre et al. [67]	An age-structured deterministic compartmental model.	Includes target groups including health workers, young people and older adults as well as mass vaccination to explore the effectiveness of vaccine.	Results show that health worker vaccination is necessary for health system resilience. Mass vaccination which includes 66% of the NSW population can achieve the herd immunity. Slower rates of vaccination can lead to COVID-19 longer, higher cases and deaths in the population.	Must be vaccinated all age group to get heard immunity.
China	Han et al. [68]	A data-driven mechanistic model with five compartments.	Seventeen age group are considered to explore the time varying vaccination effect.	A time varying vaccination program for the different age groups is the most effectively way for reducing deaths and infections. Early phase of high vaccination capacity is the key to achieve great advances of policies arrangements.	To minimize the number of deaths and ICU admissions, over 65 years older people and near of them should be vaccinated before moving to other groups.
UK	Moore et al. [69]	A modified SEIR-type model with force of infection determines by age dependent social contact matrices.	New secondary infections increase due to the first infections within a household. Secondary household contacts to be quarantined and subsequently performance no additional role.	Vaccine is most effective for elderly and vulnerable population which reduce number of deaths and healthcare demands.	To reduce death and health care demand elderly people must be vaccinated.
UK	Moore et al. [70]	Age-structured mathematical model	Incorporated two-dose vaccination and non-pharmaceutical interventions to explore the different scenarios.	vaccination alone is not sufficient to contain the outbreak of COVID-19. In the absence of non-pharmaceutical intervention, the vaccine will prevent 85% infections of the population.	Combine vaccination and non-pharmaceutical interventions is essential to eliminate COVID-19 outbreak in the UK.

Besides, to measure the optimal vaccine prioritisation of COVID-19 transmission, Han et al. (2011) developed a data-driven mechanistic model in China [68]. In this model, they considered 17 age groups and divided the population into five compartments: the unvaccinated susceptible population (S); persons who received at least the first dose of vaccine but have yet to develop protection (V); persons who received the second dose of the vaccine but failed in protection (U); infectious individuals including asymptomatic and symptomatic infections (I); and recovered or immune individuals (R). The result showed that a time-varying vaccination program for the different age groups is the most effective means of reducing deaths and infections. Furthermore, this study recommended that, to minimise the number of deaths and ICU admissions, people over 65 years of age should be vaccinated before moving to other groups such as younger and middle-aged people. Finally, the early phase of high vaccination capacity is the key to achieving significant success of policy measures and implementations [68].

Moreover, a mathematical model with different age groups in the UK was proposed by Moore et al. (2021) to investigate different COVID-19 vaccination scenarios and the age-specific vaccine efficacy [69]. A modified SEIR-type model was considered with a force of infection determined by age-dependent social contact matrices. The authors assumed that the new secondary infections increase due to the first infections within a household. However, the secondary household contacts were to be quarantined and subsequently performed no additional role for the outbreak of COVID-19. The result showed that vaccination is the most effective for the elderly and vulnerable population, which helped reduce the number of deaths and healthcare demands [69]. Modelling vaccination with non-pharmaceutical interventions is necessary to investigate significant variations in behaviours associated with COVID-19 prevention, detection and treatment than a single intervention. Furthermore, Moore et al. [70] proposed another age-structured model-integrated two-dose vaccination and non-pharmaceutical interventions in the UK. The finding showed that vaccination alone is not sufficient to contain the outbreak of COVID-19. In the absence of non-pharmaceutical interventions, the vaccine will prevent 85% of infections in the population. Combining vaccination and non-pharmaceutical interventions can eliminate the COVID-19 outbreak in the UK [70].

Statistically, modelling plays a vital role in efforts that focus on predicting, assessing, and controlling potential outbreaks of different kinds of infectious diseases. Modelling can also be used to explore the contagious disease dynamics that impact numerous variables ranging from the micro host–pathogen level to host-to-host interactions and dominant ecological, social, economic, and geographical factors across the globe. Additionally, Table 3 discusses some key literature for different infectious disease modelling approaches and their control strategies. For instance, Kanyiri et al. (2018) provide modelling results of the transmission dynamics of influenza by incorporating the aspect of drug resistance and using dynamical systems and sensitivity analysis [71]. Overall, the findings of Table 3 studies reveal some consistencies and disparities between the modelling tools and techniques, as well as the diseases and the nature of infections. Indeed, the knowledge of these modelling approaches would help develop a contemporary and robust research framework, which may specifically focus on different spatial levels within a region. Location-specific knowledge is required to develop an appropriate model for a particular area such as regional areas in NSW, Australia.

Table 3. Review of key literature for other infectious diseases modelling.

Author(s)	Research Aims	Methodology	Significant Findings
Kanyiri et al. [71]	Mathematical modelling of the transmission dynamics of influenza.	Dynamical systems, analysis of stability of stationary points, sensitivity analysis.	A mathematical model incorporating the aspect of drug resistance is formulated. The qualitative analysis of the model is given in terms of the control reproduction number, R_c. Numerical simulations reveal that despite reducing the reproduction number below unity, influenza can still persist in the population. Hence, it is essential, in addition to vaccination, to apply other strategies to curb the spread of influenza.
Wu et al. [72]	Modelling of univariate and multivariate time series data.	Transformer-based machine learning.	The authors developed a novel method which uses transformer-based machine learning models to forecast time series data. This approach works by leveraging self-attention mechanisms to learn complex patterns and dynamics from time series data. Their framework can be applied to both univariate and multivariate time series data. The authors used influenza-like illness (ILI) forecasting as a case study and showed that their transformer-based model can accurately forecast ILI prevalence using a variety of features.
Lewnard et al. [73]	Assessment of the effectiveness of interventions used in the Ebola outbreak and how these interventions may be used individually or in combination to avert future Ebola Virus Disease (EVD) outbreaks.	Building of a transmission model for the Ebola outbreak fitted to Ebola cases and deaths in Montserrado, Liberia. The model was used to assess the intervention measures such as expanding EVD treatment centres, allocation of PPE and case ascertainment numbers. 23 September 2014 was used as the base for all behaviour and contact patterns. The primary outcome measure was the expected number of cases averted by 15 December 2014.	The authors estimated that the reproductive number for EVD in Montserrado was 2.49. The allocation of 4800 additional beds at EVD treatment centres and increasing case ascertainment numbers 5-fold can avert 77,312 cases by 15 December 2014.

Table 3. *Cont.*

Author(s)	Research Aims	Methodology	Significant Findings
Kucharski et al. [74]	To understand the transmission dynamics of Zika virus (ZIKV) using a mathematical model of vector-borne infections.	A compartmental mathematical model was used to simulate vector-borne transmission. People and mosquitoes were modelled using a susceptible-exposed-infectious-removed (SEIR) framework.	An estimation of key epidemiological parameters such as the reproduction rate. Median estimates of 2.6–4.8 reproduction rates were found. An estimated 94% of the total population of the 6 archipelagos of French Polynesia were found to be infected during the outbreak. Based on the demography of French Polynesia and the results, an implication was that an initial ZIKV infection provided protection against future infections. It would also take between 12–20 years before there was a sufficient number of susceptible individuals for ZIKV to re-emerge.
Farah et al. [75]	To develop an efficient, computationally inexpensive Bayesian dynamic model for influenza.	A statistical model that combines a Gaussian process (GP) for the output function of the simulator with a dynamic linear model (DLM) for its evolution through time was developed.	The modelling framework is found to be both flexible and tractable, resulting in efficient posterior inference for the parameters of the influenza epidemic.
Luksza and Lassig [76]	To build a model to predict the evolution of the influenza virus for vaccine selection.	Sequence data which contain HA (a particular type of protein) were used to build genealogical trees. Strain frequencies were then estimated, and mutations were mapped. Predictions were done based on the model fitted. Based on the results, a vaccine strain was selected.	Factors that determine the fitness of a strain were found. A principled method for vaccine selection was suggested.
Agusto and Khan [77]	To investigate the optimal control strategy for curtailing the spread of dengue disease in Pakistan.	Optimal control theory is used to compare the different intervention strategies, including insecticide use and vaccination.	The results show that a strong reciprocal relationship exists between the insecticide use and vaccination. The cost of insecticide increases as the use of vaccination increases. Due to the increase in cost, the use of insecticide slightly increases when vaccination decreases.
Kuddus et al. [78]	To estimate the drug-resistant tuberculosis amplification rate and intervention strategies in Bangladesh.	Optimal control strategy is used to evaluate the cost-effectiveness of varying combinations of four basic control strategies—distancing, latent case finding, case holding and active case finding.	The results reveal that a combination of one or more intervention strategies is the most cost-effective way for controlling the outbreak of drug-susceptible and multi-drug resistant tuberculosis in Bangladesh.

Table 3. *Cont.*

Author(s)	Research Aims	Methodology	Significant Findings
Rahman and Kuddus [79]	To support the National Malaria Control Program for the design and characterisation of the malaria disease in Bangladesh.	A reliable qualitative and quantitative modelling technique used to identify the most influential factors in the outbreak of malaria.	From a qualitative viewpoint, the results show that service factors, disease related factors, environmental factors, and sociological factors are significant. From the quantitative modelling approach, the results reveal that the transmission rate is the most important risk factor for the outbreak of malaria in Bangladesh.
Bhunu et al. [80]	To assess the effects of smoking on the transmission dynamics of tuberculosis.	A transmission dynamics of tuberculosis model was used, considering the fact that some people in the population are smoking in order to assess the influence of smoking on tuberculosis transmission.	The results reveal that smoking enhances tuberculosis transmission and progression from latent tuberculosis cases to active tuberculosis cases. This study also shows that the number of active tuberculosis cases increases as the number of smokers increases.

3. Developing Models with a Regional Focus

COVID-19 poses a significant challenge for the government healthcare system in regional areas of NSW. One of the most significant challenges is the demand for hospitals to treat critically ill COVID-19 patients [60]. Current knowledge from the outbreak in Italy suggests that a severe demand for intensive care support can occur at the peak of an epidemic. The shortage of intensive care support often leads to preventable deaths due to the lack of accessible intensive care units (ICU) and healthcare workers [81]. The epidemic trajectory of COVID-19 in NSW seems delayed by many weeks compared to several states, including Victoria, due to the travel bans implemented at the beginning of the epidemic. The situation is changing very quickly, and NSW government policy has recently focussed on prevention rather than lockdowns or eliminating COVID-19 infection from the community [60]. Nonetheless, unless an effective vaccine is produced, it seems possible that the outbreak of this disease will transmit quickly within the general population [82]. The effectiveness of current and prospective non-pharmaceutical intervention strategies, including social distancing, is unpredictable or highly reliant on the extent to which they are implemented.

Mathematical modelling is one of the most effective ways to gain insights into the dynamics of an epidemic and to assist in the allocation of resources, including intensive care resources, during different stages of the pandemic. Fox et al. [60] developed a modified SEIR model to estimate hospitalised cases and ICU cases per 100,000 population in NSW. This study considers two scenarios; one is no intervention within a basic reproduction number of 2.4, and the other is social distancing strategies leading to a basic reproduction number of 1.6. The results showed that without social distancing measures, the peak of the COVID-19 cases for hospitalisation would be 450 per 100,000, with about 150 people needing intensive care. According to the scenario without intervention, the outbreak infection peak would be late June and hospital usage in early July. Under the second scenario with social distancing, around 180 people would be hospitalised per 100,000, with 65 people needing intensive care. The outbreak will move to early October, and peak ICU usage will move to mid-November. The authors suggested that the social distancing intervention strategy would be partially effective for the delay of the epidemic peak by around 12 weeks. However, this study did not estimate the effect of suppression strategies, which would reduce the peak of ICU demand. Therefore, further modelling is required to explore the impact of suppression strategies at the time of the epidemic in NSW, including on ICU demand. Such modelling strategies will help to notify the public concerning the timing, severity, and continuation of mitigation policies.

Weather variables including temperature, humidity and rainfall are critical determinants for the outbreak of COVID-19 in NSW [83] and other states and countries [84]. To explore the association between meteorological variables and the number of COVID-19 cases, Ward et al. [83] used a time series analysis in NSW. They used a multivariate generalised additive model (GAM) where a correlation matrix was used to select a weather variable to avoid multicollinearity in the analysis. The best model was selected based on the backward algorithm and the Akaike information criteria (AIC) value. Weather variables were analysed through a 14-day interval based on the incubation time, and the natural splines function with two degrees of freedom is used for the model trend and seasonality. The results showed that temperature and rainfall have no relationship with COVID-19 in NSW, while low temperature and low humidity are suitable for the survival and spread of the virus, because they dry out the mucous membrane, reduce the function of cilia and facilitate the spread of suspended matter in the atmosphere [84,85]. Some modelling studies suggested that lower temperatures may increase the number of COVID-19 cases [84,86]. Therefore, more research is needed to explore the association between temperature and the number of COVID-19 cases.

In the future, we propose to develop a comprehensive model of COVID-19 transmission dynamics over time to infer the impact of mitigation, suppression and multiple interventions and their cost-effective analysis for controlling COVID-19 outbreaks in NSW. We will develop a modified SEIR model to account for the following mutually exclusive compartments: Susceptible $S(t)$, uninfected individuals who are susceptible to the COVID-19 infection; Exposed $E(t)$, representing those who are infected and have not yet developed active COVID-19; Infectious $I(t)$, comprising individuals with active COVID-19; the Recovered $R(t)$, who were previously infected and successfully treated, or death $D(t)$. For estimating healthcare needs, we will categorise the infectious group into two sub-cases: Mild $M(t)$ and Critical $C(t)$; where Mild cases do not require hospital beds; and Critical cases need hospital beds. A flow diagram of our proposed model is presented in Figure 1.

To the best of our knowledge, in previous modelling studies, many mathematical models have been investigated, focusing on mysterious transmission dynamics of COVID-19 using different types of intervention strategies. However, none of them have used a cost-effective analysis for the economy in NSW, Australia. This model will consider a set of non-linear differential equations and will distinguish two essential features—the direct link between the Exposed and Recovered population and the practical healthcare demand resulting from the separation of infections into mild and critical cases. First, we will use a next-generation matrix to determine the basic reproduction number R_0 of COVID-19, where R_0 is the estimated number of secondary cases produced by single infectious cases and exclusively the susceptible population. Then, to supplement and validate the model structure, we will calibrate the number of cases from the COVID-19 data in NSW. Following this, we will perform a sensitivity analysis to explore the impact of parameters on the model outcomes. Finally, we will incorporate the economic compartment into our proposed model to explore the financial consequences of different interventions and their impact on the dynamics of COVID-19 in NSW, Australia.

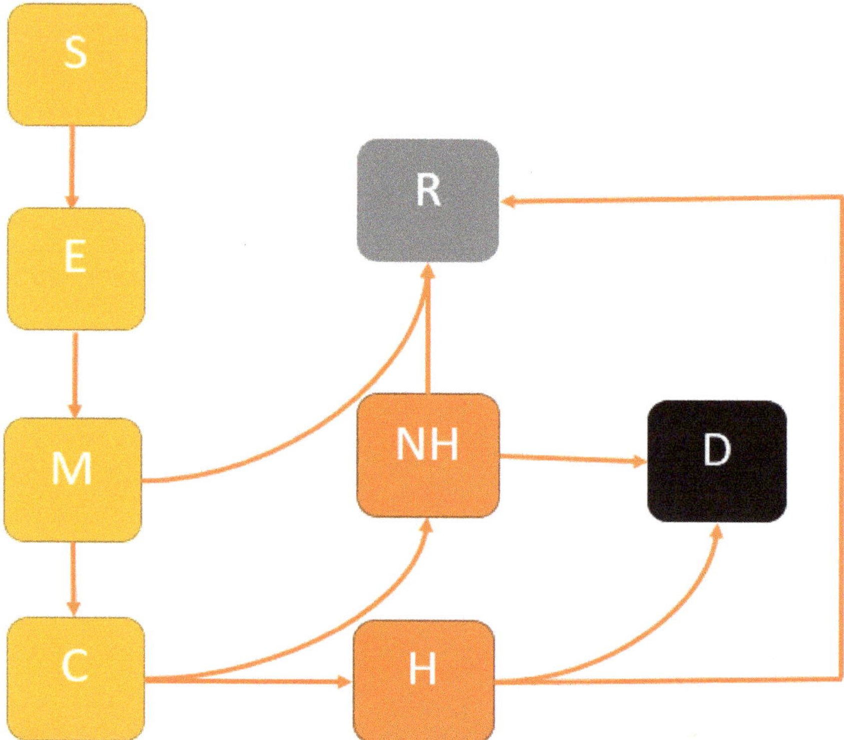

Figure 1. Extended SEIR model structure: The population is divided into the following six classes: susceptible, exposed (and not yet symptomatic), infectious (symptomatic), i.e., mild (mild or moderate symptom) and critical (severe symptoms), death and recovered (i.e., isolation, recovered, or otherwise non-infectious).

4. Conclusions

COVID-19 has had more attention from the government and media than any previous infectious disease, including influenza. Modelling studies can contribute to developing novel control methods, improving computational tools, and public data sharing. For example, modelling studies strongly advised border closures, and China first imposed an internal travel lockdown on Wuhan, which delayed the epidemic peak of COVID-19 within China but had a more significant impact on other countries [35,87,88]. Statistical modelling has also projected the shifting of outbreaks from one country to another, based on these locations' connectedness [89].

Age is a significant risk factor that can increase the severity of the outbreak of COVID-19. Mixing models can examine age-specific contact patterns and infection risk and use relative infectiousness [90]. Modelling studies have found that children are less likely to acquire an infection, and when infected, they are much less likely to show symptoms. This information will assist policymakers in strategy development. In addition, mixing models have showed that school lockdowns have a modest impact on COVID-19 transmission, encouraging authorities to re-open schools or to avoid school lockdowns completely [34,91].

Mathematical models can estimate the potential epidemic outbreak of COVID-19. One of the essential components for the modelling studies is the basic reproduction number, which is the expected secondary cases caused by a single infectious case in a susceptible population. Modelling studies have shown that implementing suppression (i.e., immediate lockdown) strategies will decrease the reproduction number to less than one, which means that the disease dies out gradually without the need to take any further action.

Furthermore, any deficiencies in performing mitigation strategies will increase the risk of having a reproduction number greater than one, which indicates that the disease persists in the population, and governments need to take more actions to control the disease [92]. Intervention strategies and government-imposed constraints on human migration have started to decrease the spread. Models presenting variations in transmission rates over time have been influential tools, helping decision-makers to implement improvements in outbreak control within public health strategies [93].

It is well known that vaccines are very effective for infectious disease control [94,95]. Therefore, for the elimination of COVID-19, a vaccine is urgently needed for global-scale use. There are many clinical trials of COVID-19 vaccines underway, though a few countries claimed success in efficacy trials at their local or national scale. Modelling studies are beneficial in evaluating the effectiveness of vaccines within clinical trials and for reducing biases [96,97]. Modelling can also assist in evaluating the possible effectiveness of vaccination policies, including location-specific ring-vaccination, age-specific vaccination, and the socioeconomic and geopolitical advantages of vaccination. However, for COVID-19, the situation is even more challenging as the disease affects different age groups differently. There is also a greater risk of co-infection and mortality with other diseases, especially in the older age group.

The information generated from the models of the COVID-19 pandemic allows collaborative involvement between decision-makers and researchers. Policymakers can provide researchers with a clear outlook of the policy settings, while researchers can construct models that assist in decision-making. Decision-makers can then plan the policy aims and the intervention strategies and should ideally build a setting where decision-makers and modellers work in combination on an ongoing basis.

Modelling studies may also perform a crucial role in expanding the scope of limited resources under discussion. For instance, a modelling study infers that UK health officials did not examine a policy that included testing due to a limited testing capacity [98]. Modellers also advise using suppression strategies in China rather than mitigation, as the results reduce exposure in China and reduce the number of global cases [99]. Modelling may also provide more optimal scenarios for different intervention strategies with significant benefits at a low cost. For example, in Australia, mitigation strategies are commonly considered rather than suppression strategies (except in Melbourne recently during the second wave of COVID-19 outbreaks) [61]. If modelling studies show that suppression strategies would provide better results, these actions can be implemented early in Australia, including in NSW. Our future application paper will consider this in the context of analysing epidemiological surveillance data to develop an optimal strategy to control COVID-19-type outbreaks in urban and rural health districts of NSW efficiently.

Non-pharmaceutical interventions and vaccination strategies are implemented to prevent and control COVID-19 in most countries in the world. Modelling can assess the potential impact of different interventions measures for mitigating the burden of COVID-19 across the globe [100]. Vaccination is the best way to prevent or control outbreaks of COVID-19. Mathematical models can examine the impact of vaccination on death if herd immunity is not achieved, and it also explores the direct effects of vaccination on reducing death are very good for which vaccines. Therefore, the steps for future research in modelling will be models with a combination of control strategies.

In this review, we have discussed some important COVID-19 models and have attempted to classify them by their structures (including some core assumptions). In addition, we summarise the model outcomes and distinctive features, including the impact of different intervention strategies and their cost, stability, and sensitivity analysis to identify the most impelling risk factors addressing model biases. In doing so, we have identified some open challenges and encouraging prospects for upcoming COVID-19 modelling-related research.

Finally, every study has its limitations. For future research, it is prudent to note those limitations that have posed a challenge to the findings of this study. This study's specific limitation is the reliance on previously published research regarding mathematical modelling of COVID-19 in three countries, including Australia, China, and the UK, from 2019 to 2021. In addition, the quality of information obtained might not always be reliable, e.g., incidence, prevalence, health demand, etc., which may contaminate findings.

Author Contributions: A.R. planned the study, A.R. and M.A.K. analysed and prepared the manuscript. R.H.L.I. and M.B. helped in the preparation and review of the manuscript. All authors have read and agreed to the published version of the manuscript.

Funding: This paper was supported by Charles Sturt University COVID-19 Research Grant No. 57. Finally, the authors acknowledge an additional financial support provided by Charles Sturt University.

Institutional Review Board Statement: Not applicable.

Informed Consent Statement: Not applicable.

Data Availability Statement: Not applicable.

Acknowledgments: The authors would like to thank Dmitry Demskoy for providing some feedback on an earlier version and colleagues in the Research Office at Charles Sturt University for their excellent services. We also acknowledge Charles Sturt University's Institute for Land, Water and Society (ILWS) for providing financial support to enable the publication of this paper.

Conflicts of Interest: The authors declare that they have no competing interests.

References

1. Harbeck, M.; Seifert, L.; Hänsch, S.; Wagner, D.; Birdsell, D.; Parise, K.; Wiechmann, I.; Grupe, G.; Thomas, A.; Keim, P.; et al. Yersinia pestis DNA from skeletal remains from the 6th century AD reveals insights into Justinianic Plague. *PLoS Pathog.* **2013**, *9*, e1003349. [CrossRef] [PubMed]
2. World Health Organization (WHO). Middle East Respiratory Syndrome Coronavirus (MERS-CoV). Available online: https://www.who.int/news-room/fact-sheets/detail/middle-east-respiratory-syndrome-coronavirus-(mers-cov) (accessed on 28 October 2021).
3. Li, Q.; Guan, X.; Wu, P.; Wang, X. Early transmission dynamics in Wuhan, China, of novel coronavirus—Infected pneumonia. *N. Engl. J. Med.* **2020**, *382*, 1199–1207. [CrossRef]
4. Fang, Y.; Nie, Y.; Penny, M. Transmission dynamics of the COVID-19 outbreak and effectiveness of government interventions: A data-driven analysis. *J. Med. Virol.* **2020**, *92*, 645–659. [CrossRef] [PubMed]
5. Raoult, D.; Zumla, A.; Locatelli, F.; Ippolito, G.; Kroemer, G. Coronavirus infections: Epidemiological, clinical and immunological features and hypotheses. *Cell Stress* **2020**, *4*, 66. [CrossRef]
6. Kang, D.; Choi, H.; Kim, J.; Choi, J. Spatial epidemic dynamics of the COVID-19 outbreak in China. *Int. J. Infect. Dis.* **2020**, *94*, 96–102. [CrossRef]
7. World Health Organization (WHO). Laboratory testing of human suspected cases of novel coronavirus (nCoV) infection. Available online: https://apps.who.int/iris/bitstream/handle/10665/330374/WHO-2019-nCoV-laboratory-2020.1-eng.pdf (accessed on 28 October 2021).
8. John Hopkins University. COVID-19 dashboard by the center for systems science and engineering. Available online: https://publichealthupdate.com/jhu/ (accessed on 28 October 2021).
9. Worldometers. Coronavirus. 2021. Available online: https://www.worldometers.info/coronavirus/ (accessed on 28 October 2021).
10. Burke, R.M.; Midgley, C.M.; Dratch, A.; Fenstersheib, M.; Haupt, T.; Holshue, M.; Ghinai, I.; Jarashow, M.C.; Lo, J.; McPherson, T.D.; et al. Active Monitoring of Persons Exposed to Patients with Confirmed COVID-19-United States, January–February 2020. *MMWR Morb. Mortal. Wkly. Rep.* **2020**, *69*, 245–246. [CrossRef]
11. Khan, S.; Ali, A.; Siddique, R.; Nabi, G. Novel coronavirus is putting the whole world on alert. *J. Hosp. Infect.* **2020**, *104*, 252–253. [CrossRef]
12. Cevik, M.; Tate, M.; Lloyd, O.; Maraolo, A.; Schafers, J.; Ho, A. SARS-CoV-2, SARS-CoV, and MERS-CoV viral load dynamics, duration of viral shedding, and infectiousness: A systematic review and meta-analysis. *Lancet Microbe* **2020**, *2*, e13–e22. [CrossRef]
13. Ghinai, I.; McPherson, T.; Hunter, J.; Kirking, H.; Christiansen, D.; Joshi, K.; Rubin, R.; Morales-Estrada, S.; Black, S.; Pacilli, M.; et al. First known person-to-person transmission of severe acute respiratory syndrome coronavirus 2 (SARS-CoV-2) in the USA. *Lancet* **2020**, *395*, 1137–1144. [CrossRef]
14. McBryde, E. The value of early transmission dynamic studies in emerging infectious diseases. *Lancet Infect. Dis.* **2020**, *20*, 512–513. [CrossRef]
15. Hui, D.S.; Azhar, E.; Kim, Y.; Memish, Z.; Oh, M.; Zumla, A. Middle East respiratory syndrome coronavirus: Risk factors and determinants of primary, household, and nosocomial transmission. *Lancet Infect. Dis.* **2018**, *18*, e217–e227. [CrossRef]

16. Khan, S.; Liu, J.; Xue, M. Transmission of SARS-CoV-2, Required Developments in Research and Associated Public Health Concerns. *Front. Med.* **2020**, *7*, 310. [CrossRef]
17. Shah, V.K.; Firmal, P.; Alam, A.; Ganguly, D.; Chattopadhyay, S. Overview of Immune Response During SARS-CoV-2 Infection: Lessons From the Past. *Front. Immunol.* **2020**, *11*, 1949. [CrossRef] [PubMed]
18. Baraniuk, C. Covid-19: What do we know about airborne transmission of SARS-CoV-2? *BMJ* **2021**, *373*, n1030. [CrossRef] [PubMed]
19. Yi, Y.; Lagniton, P.; Ye, S.; Li, E.; Xu, R. COVID-19: What has been learned and to be learned about the novel coronavirus disease. *Int. J. Biol. Sci.* **2020**, *16*, 1753–1766. [CrossRef] [PubMed]
20. Wilder-Smith, A.; Chiew, C.J.; Lee, V.J. Can we contain the COVID-19 outbreak with the same measures as for SARS? *Lancet Infect. Dis.* **2020**, *20*, e102–e107. [CrossRef]
21. Wang, K.-w.; Gao, J.; Wang, H.; Wu, X.; Yuan, Q.; Guo, F.; Zhang, Z.; Cheng, Y. Epidemiology of 2019 novel coronavirus in Jiangsu Province, China after wartime control measures: A population-level retrospective study. *Travel Med. Infect. Dis.* **2020**, *35*, 101654. [CrossRef] [PubMed]
22. Borriello, A.; Master, D.; Pellegrini, A.; Rose, J. Preferences for a COVID-19 vaccine in Australia. *Vaccine* **2021**, *39*, 473–479. [CrossRef]
23. Wang, J.; Jing, R.; Lai, X.; Zhang, H.; Lyu, Y.; Knoll, M.; Fang, H. Acceptance of COVID-19 Vaccination during the COVID-19 Pandemic in China. *Vaccines* **2020**, *8*, 482. [CrossRef] [PubMed]
24. Loomba, S.; Figueiredo, A.; Piatek, S.; de Graaf, K.; Larson, H. Measuring the impact of COVID-19 vaccine misinformation on vaccination intent in the UK and USA. *Nat. Hum. Behav.* **2021**, *5*, 337–348. [CrossRef]
25. Sallam, M. COVID-19 vaccine hesitancy worldwide: A concise systematic review of vaccine acceptance rates. *Vaccines* **2021**, *9*, 160. [CrossRef] [PubMed]
26. Chard, A.N.; Cacic-Dobo, M.; Diallo, M.; Sodha, S.; Wallace, A. Routine vaccination coverage-Worldwide, 2019. *Morb. Mortal. Wkly. Rep.* **2020**, *69*, 1706–1710. [CrossRef]
27. De Lusignan, S.; Dorward, J.; Correa, A.; Jones, N.; Akinyemi, O.; Amirthalingam, G.; Andrews, N.; Byford, R.; Dabrera, G.; Elliot, A.; et al. Risk factors for SARS-CoV-2 among patients in the Oxford Royal College of General Practitioners Research and Surveillance Centre primary care network: A cross-sectional study. *Lancet Infect. Dis.* **2020**, *20*, 1034–1042. [CrossRef]
28. Smith, S.; Morbey, R.; de Lusignan, S.; Pebody, R.; Smith, G.; Elliot, A. Investigating regional variation of respiratory infections in a general practice syndromic surveillance system. *J. Public Health* **2020**, *43*, e153–e160. [CrossRef]
29. Pareek, M.; Bangash, M.; Pareek, N.; Pan, D.; Sze, S.; Minhas, J.; Hanif, W.; Khunti, K. Ethnicity and COVID-19: An urgent public health research priority. *Lancet* **2020**, *395*, 1421–1422. [CrossRef]
30. Yang, J.; Zheng, Y.; Gou, X.; Pu, K.; Chen, Z.; Guo, Q.; Ji, R.; Wang, H.; Wang, Y.; Zhou, Y. Prevalence of comorbidities and its effects in patients infected with SARS-CoV-2: A systematic review and meta-analysis. *Int. J. Infect. Dis.* **2020**, *94*, 91–95. [CrossRef]
31. Kucharski, A.J.; Russell, T.; Diamond, C.; Liu, Y.; Edmunds, J.; Funk, S.; Eggo, R. Early dynamics of transmission and control of COVID-19: A mathematical modelling study. *Lancet Infect. Dis.* **2020**, *20*, 553–558. [CrossRef]
32. Tuite, A.R.; Fisman, D.N.; Greer, A.L. Mathematical modelling of COVID-19 transmission and mitigation strategies in the population of Ontario, Canada. *CMAJ* **2020**, *192*, E497–E505. [CrossRef]
33. Panovska-Griffiths, J. Can mathematical modelling solve the current Covid-19 crisis? *BMC Public Health* **2020**, *20*, 551. [CrossRef] [PubMed]
34. Prem, K.; Liu, Y.; Russell, T.; Kucharski, A.; Eggo, R.; Davies, N. The effect of control strategies to reduce social mixing on outcomes of the COVID-19 epidemic in Wuhan, China: A modelling study. *Lancet Public Health* **2020**, *5*, e261–e270. [CrossRef]
35. Adekunle, A.; Meehan, M.; Rojas-Alvarz, D.; Trauer, J.; McBryde, E. Delaying the COVID-19 epidemic in Australia: Evaluating the effectiveness of international travel bans. Australian and New Zealand. *J. Public Health* **2020**, *44*, 257–259.
36. Rahman, A.; Kuddus, M.A. Modelling the Transmission Dynamics of COVID-19 in Six High-Burden Countries. *BioMed Res. Int.* **2020**, *2021*, 5089184.
37. Kuddus, M.A.; Rahman, A. Analysis of COVID-19 using a modified SLIR model with nonlinear incidence. *Results Phys.* **2021**, *27*, 104478. [CrossRef] [PubMed]
38. Hou, J.; Hong, J.; Ji, B.; Dong, B.; Chen, Y.; Ward, M.; Tu, W.; Jin, Z.; Hu, J.; Su, Q. Changing transmission dynamics of COVID-19 in China: A nationwide population-based piecewise mathematical modelling study. *medRxiv* **2020**. preprint. [CrossRef]
39. Panovska-Griffiths, J.; Kerr, C.; Stuart, R.; Mistry, D.; Klein, D.; Viner, R.; Bonell, C. Determining the optimal strategy for reopening schools, work and society in the UK: Balancing earlier opening and the impact of test and trace strategies with the risk of occurrence of a secondary COVID-19 pandemic wave. *medRxiv* **2020**, preprint. [CrossRef]
40. UK Government. Coronavirus (COVID-19) in the UK. 2020. Available online: https://coronavirus.data.gov.uk/ (accessed on 28 October 2021).
41. Wang, N.; Fu, Y.; Zhang, H.; Shi, H. An evaluation of mathematical models for the outbreak of COVID-19. *Precis. Clin. Med.* **2020**, *3*, 85–93. [CrossRef]
42. Stutt, R.O.; Retkute, R.; Bradley, M.; Gilligan, C.; Colvin, J. A modelling framework to assess the likely effectiveness of facemasks in combination with 'lock-down' in managing the COVID-19 pandemic. *Proc. R. Soc. A* **2020**, *476*, 20200376. [CrossRef] [PubMed]
43. *Australia Health Sector Emergency Response Plan for Novel Coronavirus (COVID-19)*; Australia Government, Department of Health: Canberra, Australia, 2020.

44. Ip, R.H.L.; Demskoi, D.; Rahman, A.; Zheng, L. Evaluation of COVID-19 mitigation policies in Australia ssing generalised space-time autoregressive intervention models. *Int. J. Environ. Res. Public Health* **2021**, *18*, 7474. [CrossRef] [PubMed]
45. Costantino, V.; Heslop, D.J.; MacIntyre, C.R. The effectiveness of full and partial travel bans against COVID-19 spread in Australia for travellers from China during and after the epidemic peak in China. *J. Travel Med.* **2020**, *27*, taaa081. [CrossRef] [PubMed]
46. Zhao, S.; Chen, H. Modeling the epidemic dynamics and control of COVID-19 outbreak in China. *Quant. Biol.* **2020**, *11*, 1–9. [CrossRef] [PubMed]
47. Liu, M.; Ning, J.; Du, Y.; Cao, J.; Zhang, D.; Wang, J.; Chen, M. Modelling the evolution trajectory of COVID-19 in Wuhan, China: Experience and suggestions. *Public Health* **2020**, *183*, 76–80. [CrossRef] [PubMed]
48. Hao, X.; Cheng, S.; Wu, D.; Wu, T.; Lin, X.; Wang, C. Reconstruction of the full transmission dynamics of COVID-19 in Wuhan. *Nature* **2020**, *584*, 420–424. [CrossRef] [PubMed]
49. Wu, J.T.; Leung, K.; Bushman, M.; Kishore, N.; Niehus, R.; Salazar, P.; Cowling, B.; Lipsitch, M.; Leung, G. Estimating clinical severity of COVID-19 from the transmission dynamics in Wuhan, China. *Nat. Med.* **2020**, *26*, 506–510. [CrossRef] [PubMed]
50. Mizumoto, K.; Kagaya, K.; Chowell, G. Effect of the Wet Market on the coronavirus disease (COVID-19) transmission dynamics in China, 2019–2020. *Int. J. Infect. Dis.* **2020**, *97*, 96–101. [CrossRef]
51. Zhang, J.; Litvinova, M.; Liang, Y.; Wang, Y.; Wang, W.; Zhao, S.; Wu, Q.; Merler, S.; Viboud, C.; Vespignani, A.; et al. Changes in contact patterns shape the dynamics of the COVID-19 outbreak in China. *Science* **2020**, *368*, 1481–1486. [CrossRef] [PubMed]
52. Pang, L.; Liu, S.; Zhang, X.; Tian, T.; Zhao, Z. Transmission dynamics and control strategies of covid-19 in Wuhan, China. *J. Biol. Syst.* **2020**, *28*, 543–560. [CrossRef]
53. Yang, P.; Qi, J.; Zhang, S.; Wang, X.; Bi, G.; Yang, Y.; Sheng, B.; Mao, X. Feasibility of Controlling COVID-19 Outbreaks in the UK by Rolling Interventions. *medRxiv* **2020**. preprint. [CrossRef]
54. Davies, N.G.; Kucharski, A.; Eggo, R.; Gimma, A.; Edmunds, W. Effects of non-pharmaceutical interventions on COVID-19 cases, deaths, and demand for hospital services in the UK: A modelling study. *Lancet Public Health* **2020**, *5*, e375–e385. [CrossRef]
55. Booton, R.D.; MacGregor, L.; Vas, L.; Looker, K.; Hyams, C.; Bright, P.; Hading, I.; Lazarus, R.; Hamiton, F.; Lawson, D.; et al. Estimating the COVID-19 epidemic trajectory and hospital capacity requirements in South West England: A mathematical modelling framework. *BMJ Open* **2021**, *11*, e041536. [CrossRef] [PubMed]
56. Rawson, T.; Brewer, T.; Veltcheva, D.; Huntingford, C.; Bonsall, M. How and when to end the COVID-19 lockdown: An optimization approach. *Front. Public Health* **2020**, *8*, 262. [CrossRef]
57. Thompson, R.N. Epidemiological models are important tools for guiding COVID-19 interventions. *BMC Med.* **2020**, *18*, 152. [CrossRef] [PubMed]
58. Peiliang, S.; Li, K. An SEIR Model for Assessment of Current COVID-19 Pandemic Situation in the UK. *medRxiv* **2020**. preprint. [CrossRef]
59. Chang, S.L.; Harding, N.; Zachreson, C.; Cliff, O.; Prokopenko, M. Modelling transmission and control of the COVID-19 pandemic in Australia. *Nat. Comm.* **2020**, *11*, 5710. [CrossRef]
60. Fox, G.J.; Trauer, J.M.; McBryde, E. Modelling the impact of COVID-19 on intensive care services in New South Wales. *Med J. Aust.* **2020**, *212*, 468–469. [CrossRef]
61. Moss, R.; Wood, J.; Brown, D.; Shearer, F.; Black, A.; Cheng, A.; McCaw, J.; McVernon, J. Modelling the impact of COVID-19 in Australia to inform transmission reducing measures and health system preparedness. *medRxiv* **2020**. preprint. [CrossRef]
62. Milne, G.J.; Xie, S. The effectiveness of social distancing in mitigating COVID-19 spread: A modelling analysis. *medRxiv* **2020**. preprint. [CrossRef]
63. Price, D.J.; Shearer, F.; Meehan, M.; McBryde, E.; Moss, R.; Golding, N.; Conway, E.; Dawson, P.; Cromer, D.; Wood, J.; et al. Early analysis of the Australian COVID-19 epidemic. *eLife* **2020**, *9*, e58785. [CrossRef]
64. Bubar, K.M.; Reinholt, K.; Kissler, S.M.; Lipsitch, M.; Cobey, S.; Grad, Y.H.; Larremore, D.B. Model-informed COVID-19 vaccine prioritization strategies by age and serostatus. *Science* **2021**, *371*, 916–921.
65. Giordano, G.; Colaneri, M.; Fillippo, A.; Blanchini, F.; Bolzern, P.; Nicolao, G.; Sacchi, P.; Colaneri, P.; Bruno, R. Modeling vaccination rollouts, SARS-CoV-2 variants and the requirement for non-pharmaceutical interventions in Italy. *Nat. Med.* **2021**, *27*, 993–998. [CrossRef] [PubMed]
66. McBryde, E.; Meehan, M.; Caldwell, J.; Adekunle, A.; Ogunlade, S.; Kuddus, M.; Ragonnet, R.; Jayasundara, P.; Trauer, J.; Cope, R. Modelling direct and herd protection effects of vaccination against the SARS-CoV-2 Delta variant in Australia. *Med. J. Aust.* **2021**, 34477236. [CrossRef]
67. MacIntyre, C.R.; Costantino, V.; Trent, M. Modelling of COVID-19 vaccination strategies and herd immunity, in scenarios of limited and full vaccine supply in NSW, Australia. *Vaccine* in press. **2021**. [CrossRef]
68. Han, S.; Cai, J.; Yang, J.; Zhang, J.; Wu, Q.; Zheng, W.; Shi, H.; Ajelli, M.; Zhou, X.; Yu, H. Time-varying optimization of COVID-19 vaccine prioritization in the context of limited vaccination capacity. *Nat. Commun.* **2021**, *12*, 4673. [CrossRef]
69. Moore, S.; Hill, E.; Dyson, L.; Tildesley, M.; Keeling, M. Modelling optimal vaccination strategy for SARS-CoV-2 in the UK. *PLoS Comput. Biol.* **2021**, *17*, e1008849. [CrossRef] [PubMed]
70. Moore, S.; Hill, E.; Tildesley, M.; Dyson, L.; Keeling, M. Vaccination and non-pharmaceutical interventions for COVID-19: A mathematical modelling study. *Lancet Infect. Dis.* **2021**, *21*, 793–802. [CrossRef]
71. Kanyiri, C.W.; Mark, K.; Luboobi, L. Mathematical Analysis of Influenza A Dynamics in the Emergence of Drug Resistance. *Comput. Math. Methods Med.* **2018**, *2018*, 2434560. [CrossRef]

72. Wu, N.; Green, B.; Ben, X.; O'Banion, S. Deep Transformer Models for Time Series Forecasting: The Influenza Prevalence Case. *arXiv* **2020**, arXiv:2001.08317.
73. Lewnard, J.A.; Ndeffo, M.; Alfaro-Murillo, J.; Altice, F.; Bawo, L.; Nyenswah, T.; Galvani, A. Dynamics and control of Ebola virus transmission in Montserrado, Liberia: A mathematical modelling analysis. *Lancet Infect. Dis.* **2014**, *14*, 1189–1195. [CrossRef]
74. Kucharski, A.J.; Funk, S.; Eggo, R.; Mallet, H.; Edmunds, W.; Nilles, E. Transmission dynamics of Zika virus in island populations: A modelling analysis of the 2013–2014 French Polynesia outbreak. *PLoS Negl. Trop. Dis.* **2016**, *10*, e0004726. [CrossRef]
75. Farah, M.; Birrell, P.; Conti, S.; Angelis, D. Bayesian emulation and calibration of a dynamic epidemic model for A/H1N1 influenza. *J. Am. Stat. Assoc.* **2014**, *109*, 1398–1411. [CrossRef]
76. Łuksza, M.; Lässig, M. A predictive fitness model for influenza. *Nature* **2014**, *507*, 57–61. [CrossRef] [PubMed]
77. Agusto, F.; Khan, M. Optimal control strategies for dengue transmission in Pakistan. *Math. Biosci.* **2018**, *305*, 102–121. [CrossRef]
78. Kuddus, M.A.; Meehan, M.; White, L.; McBryde, E.; Adekunle, A. Modeling drug-resistant tuberculosis amplification rates and intervention strategies in Bangladesh. *PLoS ONE* **2020**, *15*, e0236112. [CrossRef] [PubMed]
79. Rahman, A.; Kuddus, M.A. Cost-effective modeling of the transmission dynamics of malaria: A case study in Bangladesh. *Commun. Stat. Case Stud. Data Anal. Appl.* **2020**, *6*, 270–286. [CrossRef]
80. Bhunu, C.; Mushayabasa, S.; Tchuenche, J. A theoretical assessment of the effects of smoking on the transmission dynamics of tuberculosis. *Bull. Math. Biol.* **2011**, *73*, 1333–1357. [CrossRef]
81. Livingston, E.; Bucher, K. Coronavirus disease 2019 (COVID-19) in Italy. *JAMA* **2020**, *323*, 1335. [CrossRef]
82. Hunter, P. The spread of the COVID-19 coronavirus: Health agencies worldwide prepare for the seemingly inevitability of the COVID-19 coronavirus becoming endemic. *EMBO Rep.* **2020**, *21*, e50334. [CrossRef]
83. Ward, M.P.; Xiao, S.; Zhang, Z. The Role of Climate During the COVID-19 epidemic in New South Wales, Australia. *Transbound. Emerg. Dis.* **2020**, *67*, 2313–2317. [CrossRef] [PubMed]
84. Qi, H.; Xiao, S.; Shi, R.; Ward, M.; Chen, Y.; Tu, W.; Su, Q.; Wang, W.; Wang, X.; Zhang, Z. COVID-19 transmission in Mainland China is associated with temperature and humidity: A time-series analysis. *Sci. Total. Environ.* **2020**, *728*, 138778. [CrossRef]
85. Jamil, T.; Alam, I.; Gojobori, T.; Duarte, C. No evidence for temperature-dependence of the COVID-19 epidemic. *Front. Public Health* **2020**, *8*, 436. [CrossRef] [PubMed]
86. Shi, P.; Dong, Y.; Yan, H.; Zhao, C.; Li, X.; Liu, W.; He, M.; Tang, S.; Xi, S. Impact of temperature on the dynamics of the COVID-19 outbreak in China. *Sci. Total. Environ.* **2020**, *728*, 138890. [CrossRef] [PubMed]
87. Hou, J.; Hong, J.; Ji, B.; Dong, B.; Chen, Y.; Ward, M.; Tu, W.; Jin, Z.; Hu, J.; Su, Q.; et al. Changed transmission epidemiology of COVID-19 at early stage: A nationwide population-based piecewise mathematical modelling study. *Travel Med. Infect. Dis.* **2021**, *39*, 101918. [CrossRef]
88. Bhatia, S.; Imao, N.; Cuomo-Dannenburg, G.; Baguelin, M.; Boonyasiri, A.; Cori, A.; Cucunuba, Z.; Dorigatti, I.; Fitzjohn, R.; Fu, H.; et al. *Report 6: Relative Sensitivity of International Surveillance*; Imperial College London: London, UK, 2020.
89. Adegboye, O.; Adekunle, A.; Pak, A.; Gayawan, E.; Leung, D.; Rojas, D.; Elfaki, F.; McBryde, E.; Eisen, D. Change in outbreak epicenter and its impact on the importation risks of COVID-19 progression: A modelling study. *Travel Med. Infect. Dis.* **2021**, *40*, 101988. [CrossRef]
90. Mossong, J.; Hens, N.; Jit, M.; Beutels, P.; Auranen, K.; Mikolajczzyk, R.; Massari, M.; Salmaso, S.; Tomba, G.; Wallinga, J.; et al. Social contacts and mixing patterns relevant to the spread of infectious diseases. *PLoS Med.* **2008**, *5*, e74. [CrossRef] [PubMed]
91. McBryde, E.S.; Trauer, J.; Adekunle, A.; Ragonnet, R.; Meehan, M. Stepping out of lockdown should start with school re-openings while maintaining distancing measures. Insights from mixing matrices and mathematical models. *medRxiv* **2020**. preprint. [CrossRef]
92. Ferguson, N.; Laydon, D.; Nedjati-Gilani, G.; Imai, N.; Ainslie, K.; Baguelin, M.; Bhatia, S.; Boonyasiri, A.; Cucunuba, Z.; Cuomo-Dannenburg, G.; et al. *Report 9: Impact of Non-Pharmaceutical Interventions (NPIs) to Reduce COVID19 Mortality and Healthcare Demand*; Imperial College London: London, UK, 2020. [CrossRef]
93. Taylor, C. How New Zealand's 'eliminate' Strategy Brought New Coronavirus Cases down to Zero. CNBC. 2020. Available online: www.cnbc.com (accessed on 28 October 2021).
94. Brooks, J.V.; Frank, A.; Keen, M.; Bellisle, J.; Orme, I. Boosting vaccine for tuberculosis. *Infect. Immun.* **2001**, *69*, 2714–2717. [CrossRef] [PubMed]
95. Drape, R.J.; Macklin, M.; Barr, L.; Jones, S.; Haynes, J.; Dean, H. Epidermal DNA vaccine for influenza is immunogenic in humans. *Vaccine* **2006**, *24*, 4475–4481. [CrossRef]
96. Ragonnet, R.; Trauer, J.; Denholm, J.; Geard, N.; Hellard, M.; McBryde, E. Vaccination programs for endemic infections: Modelling real versus apparent impacts of vaccine and infection characteristics. *Sci. Rep.* **2015**, *5*, 15468. [CrossRef]
97. Wu, Y.; Marsh, J.; McBryde, E.; Snelling, T. The influence of incomplete case ascertainment on measures of vaccine efficacy. *Vaccine* **2018**, *36*, 2946–2952. [CrossRef] [PubMed]
98. Adam, D. Special report: The simulations driving the world's response to COVID-19. *Nature* **2020**, *580*, 316. [CrossRef]
99. Sridhar, D.; Majumder, M.S. Modelling the pandemic. *BMJ* **2020**, *369*, m1567. [CrossRef]
100. Abdulla, F.; Nain, Z.; Karimuzzaman, M.; Hossain, M.M.; Rahman, A. A non-linear biostatistical graphical modeling of preventive actions and healthcare factors in controlling COVID-19 pandemic. *Int. J. Environ. Res. Public Health* **2021**, *18*, 4491. [CrossRef] [PubMed]

MDPI
St. Alban-Anlage 66
4052 Basel
Switzerland
Tel. +41 61 683 77 34
Fax +41 61 302 89 18
www.mdpi.com

Viruses Editorial Office
E-mail: viruses@mdpi.com
www.mdpi.com/journal/viruses

www.ingramcontent.com/pod-product-compliance
Lightning Source LLC
LaVergne TN
LVHW070848170726
843515LV00005B/600

* 9 7 8 3 0 3 6 5 4 0 2 8 3 *